TOPICS IN HARMONIC ANALYSIS

The Appleton-Century Mathematics Series

Raymond W. Brink and John M. H. Olmsted, Editors

TOPICS IN HARMONIC ANALYSIS

Charles F. Dunkl & Donald E. Ramirez

Department of Mathematics • University of Virginia

APPLETON-CENTURY-CROFTS
EDUCATIONAL DIVISION
New York **MEREDITH CORPORATION**

761-1

Library of Congress Card Number: 73-153387

PRINTED IN THE UNITED STATES OF AMERICA

390-27819-X

PREFACE

Our purpose in writing this book was to supplement the 1962 book of Rudin, *Fourier Analysis on Groups*, for our harmonic analysis course and seminar at the University of Virginia. The first part (Chapters 1–6) is about locally compact abelian groups and it includes a complete discussion of the maximal ideal space of $M(G)$ and the new proof of the Cohen idempotent theorem by Itô and Amemiya. The second part (Chapters 7–10) is an invitation to harmonic analysis on compact non-abelian groups. It contains a discussion of the algebra $A(G)$ (the non-abelian analogue of $l^1(G)$), spherical harmonics, the Poisson integral, and analytic functions in the n-complex ball. Appendix A contains Bredon's proof of the existence and uniqueness of Haar measure. Appendix B discusses integration algebras and the Hausdorff-Young-Kunze theorem. Appendix C describes some current research on compact groups.

Our topics clearly do not include all the significant new results since 1962; for example, the work of Varopoulos on $M_0(G)$ has not been included because of space and time limitations.

We use □ to indicate the end of a proof. The symbol $\subset$ denotes containment whereas $\subsetneq$ denotes proper containment. The paragraphs which compose the text are numbered consecutively. For example, Theorem 7.2.8 is the eighth paragraph in Section 2 of Chapter 7. In Chapter 7 this theorem is referred to as Theorem 2.8. To refer to a reference book we simply invoke a symbol; for example, [R] denotes Rudin's book, *Fourier*

Analysis on Groups. To cite a particular paper of Rudin we would write, for example, Rudin [4]. We caution the reader to be aware that the definition of a Fourier-Stieltjes transform of a measure on a locally compact abelian group involves no inverse whereas the inverse is employed in the compact nonabelian definition.

Since we use a great amount of notation we have provided an index of special symbols which refers the reader to the respective definitions. The historical notes scattered throughout the book give the basic references for the various theorems. Our references are not meant to be a complete listing of all works in the field.

We wish to thank Miss Blanche Bailey for her assistance in preparing the manuscript.

The authors have been partially supported by NSF contract number GP-8981.

C.F.D.
D.E.R.

TABLE OF CONTENTS

Chapter 4: WEAKLY ALMOST PERIODIC FUNCTIONS

Chapter 5: SIDON SETS

Chapter 6: IDEMPOTENT MEASURES

Chapter 7: INTRODUCTION TO COMPACT GROUPS

Chapter 8: BANACH SPACES ON THE DUAL OF A COMPACT GROUP

TOPICS
IN
HARMONIC ANALYSIS

CHAPTER 1

THE MAXIMAL IDEAL SPACE OF M(G)

1. Introduction

1.1: Let G be a locally compact abelian (LCA) group. Let $C_0(G)$ denote the space of complex-valued continuous functions on G vanishing at infinity and $M(G)$ the space of all finite regular complex Borel measures on G. Let us note how $M(G)$, the conjugate space of $C_0(G)$, can be represented as a semigroup of bounded operators on $C_0(G)$. For each $\mu \in M(G)$, define E_μ by the following commutative diagram:

$$\begin{array}{ccccc} & C_0(G) & \xrightarrow{E_\mu} & C_0(G) & \\ R(x) & \downarrow & & \downarrow & e_x \\ & C_0(G) & \xrightarrow{\mu} & \mathbf{C} & \end{array}$$

where, for $x \in G$, $R(x)$ denotes translation by x, $R(x)f(y) = f(y + x)$, $f \in C_0(G)$, $y \in G$, and e_x denotes evaluation at x. Convolution of measures can be defined by the rule

$$\mu * \nu(f) = \mu(E_\nu(f))$$

where $f \in C_0(G)$. Thus $E_{\mu*\nu} = E_\mu E_\nu$; that is, we induce in $M(G)$ the multiplication associated with the composition of the corresponding operators. In the same way we show that $M(G)^*$, the conjugate space of $M(G)$, can be represented as a semigroup of bounded operators on $M(G)$ [Theorem 2.3]. This representation is computable [Lemma 2.6], and one shows that $M(G)^*$ with this multiplication is a commutative B^*-algebra with identity [Theorem 2.9]. Let Δ denote the space of nonzero multiplicative linear functionals on $M(G)$, and P the norm-closed linear span of Δ in $M(G)^*$. Now P is a commutative B^*-subalgebra with identity of $M(G)^*$ [Theorem 2.10]. Thus $P \cong C(B)$, the algebra of all complex-valued continuous functions on a compact Hausdorff space B. On B we introduce a semigroup operation such that the set $\check{B}$ of nonzero semicharacters on B is identified with Δ [Theorem 3.5]. We call B the structure semigroup for $M(G)$. In $\check{B}$, let $\hat{I}$ denote the semicharacters whose absolute values assume only the values 0 or 1. We then show that $\hat{I}$ is dense in the Shilov boundary of $M(G)$ [Theorems 3.8 and 3.13]. This result will be needed in Chapter 2. Finally, we show how Δ can be represented by generalized characters.

We will denote the character group of G by Γ or $\hat{G}$, and $M(G)\hat{}$ the algebra of Fourier-Stieltjes transforms: $\mu\,(\gamma) = \int_G \gamma(x)\, d\mu(x), \gamma \in \Gamma$.

2. The Operator Algebra M(G)*

2.1 Definition: Let G be a LCA group. For $F \in M(G)^*$ and $\mu \in M(G)$, define the complex-valued function, $(E_F\mu)\hat{}$, on $\Gamma(= \hat{G})$ by

$$(E_F\mu)\hat{}(\gamma) = F(\gamma\, d\mu), \qquad \gamma \in \Gamma$$

(Lemma 2.2 will justify our notation).

2.2 Lemma: *$(E_F\mu)\hat{} \in M(G)\hat{}$ and $\mu \mapsto E_F\mu$ is a bounded linear operator on $M(G)$, with $\| E_F \| = \| F \|$.*

Proof: We first note that if $\gamma_\alpha \xrightarrow{\alpha} \gamma$ in Γ (in the compact-open topology) then $\gamma_\alpha\, d\mu \xrightarrow{\alpha} \gamma\, d\mu$ in $M(G)$, and so $F(\gamma_\alpha\, d\mu) \xrightarrow{\alpha} F(\gamma\, d\mu)$. Thus $(E_F\mu)\hat{}$ is a continuous function on Γ.

For $c_1, \ldots, c_n \in \mathbf{C}$ and $\gamma_1, \ldots, \gamma_n \in \Gamma$,

$$\left| \sum_{i=1}^{n} c_i (E_F\mu)\hat{}(\gamma_i) \right| = \left| \sum_{i=1}^{n} c_i F(\gamma_i\, d\mu) \right|$$

$$\leq \| F \| \left\| \sum_{i=1}^{n} c_i \gamma_i\, d\mu \right\| \leq \| F \| \, \| \mu \| \sup \left\{ \left| \sum_{i=1}^{n} c_i \gamma_i(x) \right| : x \in G \right\}.$$

Thus by Eberlein's theorem [R, p. 32],

$$(E_F\mu)\hat{} \in M(G)\hat{} \text{ and } \| E_F\mu \| \leq \| F \| \, \| \mu \| .$$

Thus $\| E_F \| \leq \| F \|$.

Now

$$| F(\mu) | = | (E_F\mu)\hat{}(0) | \leq \| E_F\mu \| \leq \| E_F \| \, \| \mu \| .$$

Thus $\| F \| = \| E_F \|$. □

2.3 Theorem: *The mapping $F \mapsto E_F$ is a one-to-one, onto, linear isometry between $M(G)^*$ and $\mathscr{B}$, the semigroup of bounded operators on $M(G)$ which commute with translation by $\gamma \in \Gamma$; that is for $E \in \mathscr{B}$, $\gamma\, dE(\mu) = E(\gamma\, d\mu)$, $\mu \in M(G)$, $\gamma \in \Gamma$.*

Proof: We have by the previous lemma that $F \mapsto E_F$ is a linear isometry from $M(G)^*$ into $\mathscr{B}(M(G))$, the bounded operators on $M(G)$. Now E_F commutes with translation by $\gamma \in \Gamma$ since for $\gamma_1, \gamma_2 \in \Gamma$,

$$(\gamma_1\, dE_F\mu)\hat{}(\gamma_2) = (E_F\mu)\hat{}(\gamma_1 + \gamma_2) = F(\gamma_1\gamma_2\, d\mu) = E_F(\gamma_1\, d\mu)\hat{}(\gamma_2),$$

[see 3.1.3.e].

For $E \in \mathscr{B}$, define $F \in M(G)^*$ by $F(\mu) = (E\mu)\hat{}(0)$, $\mu \in M(G)$. Now $E_F = E$ since

$$(E_F\mu)\hat{}(\gamma) = F(\gamma\, d\mu) = E(\gamma\, d\mu)\hat{}(0) = (\gamma\, dE\mu)\hat{}(0) = (E\mu)\hat{}(\gamma),$$

for $\mu \in M(G)$ and $\gamma \in \Gamma$. □

2.4 Corollary: *Let Δ denote the nonzero multiplicative linear functionals on $M(G)$ and $\mathscr{E}$ the nonzero algebraic endomorphisms in $\mathscr{B}$. The map $\pi \mapsto E_\pi$ is a one-to-one, onto, linear isometry between Δ and the semigroup $\mathscr{E}$.*

Proof: Let $\pi \in \Delta$, then E_π is multiplicative since

$$\begin{aligned}(E_\pi(\mu * \nu))\hat{}(\gamma) &= \pi(\gamma\, d\mu * \nu) = \pi(\gamma\, d\mu * \gamma\, d\nu) = \pi(\gamma\, d\mu)\pi(\gamma\, d\nu)\\ &= (E_\pi\mu)\hat{}(\gamma)\,(E_\pi\nu)\hat{}(\gamma),\end{aligned}$$

for $\gamma \in \Gamma$.

If $E \in \mathscr{E}$, then the linear functional π defined by $\pi(\mu) = (E\mu)\hat{}(0)$ is multiplicative and nonzero and hence in Δ.

Finally, if $\pi_1, \pi_2 \in \Delta$, then $E_{\pi_1} \circ E_{\pi_2}$ is nonzero since, letting δ be the unit in $M(G)$, $E_\pi(\delta) = \delta((E_\pi(\delta))\hat{}(\gamma) = \pi(\gamma\, d\delta) = \pi(\delta) = 1)$. □

2.5 Definition: Let $M^+ = \{\mu \in M(G) : \mu \geq 0\}$. For $\mu \in M^+$ and $F \in M(G)^*$, let $f_F^\mu \in L^1(\mu)^* = L^\infty(\mu)$ be defined by restricting F to $L^1(\mu)$; that is, $f_F^\mu = F \,|\, L^1(\mu)$. We consider f_F^μ as an element of $L^\infty(\mu)$.

2.6 Lemma: *Let $\mu \in M^+$. For $\nu \in L^1(\mu)$, $E_F\nu = f_F^\mu\, d\nu$.*

Proof: Note that

$$\int_G \gamma\, dE_F\nu = F(\gamma\, d\nu) = \int_G f_F^\mu \gamma\, d\nu = \int_G \gamma f_F^\mu\, d\nu.$$

Thus $(E_F\nu)\hat{} = (f_F^\mu\, d\nu)\hat{}$. □

2.7 Definition: For $F, H \in M(G)^*$, we define $F \times H \in M(G)^*$ by $E_{F \times H} = E_F E_H$; that is, we induce from the natural semigroup operation in $\mathscr{B}$ a semigroup operation, $\times$, on $M(G)^*$ via the identification from Theorem 2.3. Note that for $\mu \in M^+$, $\nu \ll \mu$, and $F, H \in M(G)^*$ we have that $E_{F \times H}\nu = f_F^\mu f_H^\mu \, d\nu$.

For $F \in M(G)^*$, we define $F^\sim \in M(G)^*$ by $F^\sim(\mu) = \overline{F(\bar{\mu})}$, where $\bar{\mu}(f) = \overline{\mu(\bar{f})}$ for $\mu \in M(G)$ and $f \in C_0(G)$.

2.8 Lemma: *Let $\mu \in M^+$. For $\nu \in L^1(\mu)$, $E_{F^\sim}\nu = \overline{f_F^\mu}\, d\nu$.*

Proof: We first note that

$$\int_G \gamma \, dE_{F^\sim}\nu = F^\sim(\gamma \, d\nu) = \overline{F(\bar{\gamma}\, d\bar{\nu})} = \overline{\int_G f_F^\mu \bar{\gamma}\, d\bar{\nu}}$$
$$= \int_G \overline{f_F^\mu} \gamma \, d\nu = \int_G \gamma \overline{f_F^\mu}\, d\nu.$$

As in Lemma 2.6, this yields that $E_{F^\sim}\nu = \overline{f_F^\mu}\, d\nu$. □

2.9 Theorem: *$M(G)^*$ with the operation $\times$ is a commutative B^*-algebra with identity.*

Proof: Let $F \in M(G)^*$. We first show that $\| F^\sim \times F \| = \| F \|^2$. This follows from

$$\| F^\sim \times F \| = \| E_{F^\sim \times F} \| = \| E_{F^\sim} E_F \|$$
$$= \sup\left\{\frac{\| E_{F^\sim} E_F \nu \|}{\| \nu \|} : \mu \in M^+, \nu \in L^1(\mu), \nu \neq 0\right\}$$
$$= \sup\left\{\frac{\| \overline{f_F^\mu} f_F^\mu \, d\nu \|}{\| \nu \|} : \mu \in M^+, \nu \in L^1(\mu), \nu \neq 0\right\}$$
$$= \sup\{\| \overline{f_F^\mu} f_F^\mu \|_\infty : \mu \in M^+\} = \sup\{\| f_F^\mu \|_\infty^2 : \mu \in M^+\}$$
$$= (\sup\{\| f_F^\mu \|_\infty : \mu \in M^+\})^2$$
$$= \left(\sup\left\{\frac{\| f_F^\mu \, d\nu \|}{\| \nu \|} : \mu \in M^+, \nu \in L^1(\mu), \nu \neq 0\right\}\right)^2$$
$$= \| E_F \|^2 = \| F \|^2.$$

Hence $(M(G)^*, \times)$ is a B^*-algebra.

$M(G)^*$ is commutative since for $\mu \in M^+$, $\nu \in L^1(\mu)$, we have

$$(F \times H)(\nu) = (E_{F \times H}\nu)\hat{}(0) = (E_F E_H \nu)\hat{}(0)$$
$$= (f_F^\mu f_H^\mu \, d\nu)\hat{}(0) = (f_H^\mu f_F^\mu \, d\nu)\hat{}(0)$$
$$= (E_H E_F \nu)\hat{}(0) = (E_{H \times F}\nu)\hat{}(0) = (H \times F)(\nu).$$

Let $\pi_0 \in \Delta$ be defined by $\pi_0(\lambda) = \hat{\lambda}(0) = \int_G 1 \, d\lambda$, where $\lambda \in M(G)$. For $\mu \in M^+, f_{\pi_0}^\mu = 1$ and so π_0 is the identity in $M(G)^*$. In summary, $M(G)^*$ is a commutative B^*-algebra with identity. □

2.10 Theorem: *Let Δ denote the set of nonzero multiplicative linear functionals on $M(G)$ and let P be the norm-closed linear span of Δ in $M(G)^*$; that is, $P = \overline{Sp\Delta} \subset M(G)^*$. With the induced operations on P as a subspace of $M(G)^*$, P is a commutative B^*-algebra with identity.*

Proof: Δ is closed under $\tilde{}$ since for $\pi \in \Delta$ and $\mu, \nu \in M(G)$,

$$\tilde{\pi}(\mu * \nu) = \overline{\pi(\overline{\mu * \nu})} = \overline{\pi(\bar{\mu} * \bar{\nu})} = \overline{\pi(\bar{\mu})}\,\overline{\pi(\bar{\nu})} = \tilde{\pi}(\mu)\tilde{\pi}(\nu).$$

That Δ is closed under $\times$ follows since $\mathscr{E}$ is closed under composition by Corollary 2.4. □

3. The Structure Semigroup

3.1 Definition: With the notation from Theorem 2.10, we have that P is a commutative B^*-algebra with identity and thus $P \cong C(B)$, where B is the maximal ideal space of P, and thus B is a compact Hausdorff space. We call B the **structure semigroup** for $M(G)$.

3.2 Notation: For $\pi \in \Delta \subset P$, let $\hat{\pi} \in C(B)$ denote the Gelfand representation of π. Define $\alpha: \hat{\Delta} \subset C(B) \to C(B \times B)$ by $(\alpha\hat{\pi})(s, t) = \hat{\pi}(s)\hat{\pi}(t)$, where $s, t \in B$. Since the Gelfand representation of $M(G)$ strongly separates the points of Δ, Δ is linearly independent in $M(G)^*$; and therefore we may extend α linearly to $Sp\hat{\Delta}$.

Let δ_s be the unit point mass at $s \in B$ and $\delta_{(s,t)}$ be the unit point mass at $(s, t) \in B \times B$. Now for $a_i \in \mathbf{C}$, $\pi_i \in \Delta$, $1 \le i \le n$, we have

$$\begin{aligned}
&|\delta_{(s,t)} \circ \alpha(a_1\hat{\pi}_1 + \cdots + a_n\hat{\pi}_n)| \\
&\quad = |a_1\hat{\pi}_1(s)\hat{\pi}_1(t) + \cdots + a_n\hat{\pi}_n(s)\hat{\pi}_n(t)| \\
&\quad = |\delta_s(a_1\hat{\pi}_1(t)\hat{\pi}_1 + \cdots + a_n\hat{\pi}_n(t)\hat{\pi}_n)| \\
&\quad \le \|\delta_s\|\,\|a_1\hat{\pi}_1(t)\hat{\pi}_1 + \cdots + a_n\hat{\pi}_n(t)\hat{\pi}_n\|_{C(B)} \\
&\quad = \|a_1\hat{\pi}_1(t)\pi_1 + \cdots + a_n\hat{\pi}_n(t)\pi_n\|_P \\
&\quad = \|a_1\hat{\pi}_1(t)\pi_1 + \cdots + a_n\hat{\pi}_n(t)\pi_n\|_{M(G)^*} \\
&\quad = \sup_\mu |a_1\hat{\pi}_1(t)\pi_1(\mu) + \cdots + a_n\hat{\pi}_n(t)\pi_n(\mu)| \\
&\quad = \sup_\mu |\delta_t(a_1\pi_1(\mu)\hat{\pi}_1 + \cdots + a_n\pi_n(\mu)\hat{\pi}_n)| \\
&\quad \le \|\delta_t\| \sup_\mu \|a_1\pi_1(\mu)\hat{\pi}_1 + \cdots + a_n\pi_n(\mu)\hat{\pi}_n\|_{C(B)} \\
&\quad = \sup_\mu \|a_1\pi_1(\mu)\pi_1 + \cdots + a_n\pi_n(\mu)\pi_n\|_{M(G)^*}
\end{aligned}$$

$$
\begin{aligned}
&= \sup_{\mu} \sup_{\nu} | a_1\pi_1(\mu)\pi_1(\nu) + \cdots + a_n\pi_n(\mu)\pi_n(\nu) | \\
&= \sup_{\mu} \sup_{\nu} | a_1\pi_1(\mu * \nu) + \cdots + a_n\pi_n(\mu * \nu) | \\
&\leq \sup_{\mu} \sup_{\nu} \| a_1\pi_1 + \cdots + a_n\pi_n \|_{M(G)^*} \| \mu * \nu \|_{M(G)} \\
&\leq \| a_1\pi_1 + \cdots + a_n\pi_n \|_{M(G)^*} \\
&= \| a_1\hat{\pi}_1 + \cdots + a_n\hat{\pi}_n \|_{C(B)},
\end{aligned}
$$

where, for example, $\sup_\mu$ is the supremum over all elements μ with $\| \mu \|_{M(G)} \leq 1$. Thus, α is bounded on $Sp\Delta$ and may be extended to all of $C(B)$. Call the extension β. Thus, $\beta: C(B) \to C(B \times B)$, $\| \beta \| = 1$, and $(\beta\hat{\pi})(s, t) = \hat{\pi}(s)\hat{\pi}(t)$, $\hat{\pi} \in \hat{\Delta}$.

Using the map β, we define a natural multiplication in B. The map $\delta_{(s,t)} \circ \beta: C(B) \to \mathbf{C}$ is a nonzero multiplicative linear functional, and thus there is $r \in B$ such that $\delta_{(s,t)} \circ \beta = \delta_r$. Define $m: B \times B \to B$ by $m(s, t) = r$. We write st for $m(s, t)$.

3.3 Theorem: *B is a compact abelian topological semigroup, and $M(G)$ is isometric and isomorphic to a weak-$*$ dense subalgebra of $M(B)$.*

Proof: We first show that $m: B \times B \to B$ is continuous. Let $V = \{s \in B: |f(s)| < \delta\}$ for some $f \in C(B)$ and $\delta > 0$. Now V is a typical subbasic neighborhood in B. Let $U = \{(s, t) \in B \times B: |\beta f(s, t)| < \delta\}$. Thus, U is a neighborhood in $B \times B$ and $m(U) \subset V$ since $\beta f(s, t) = f(m(s, t)) = f(st)$.

If $\hat{\pi}(s) = \hat{\pi}(t)$, $s, t \in B$, for all $\hat{\pi} \in \hat{\Delta}$, then $s = t$, since $Sp\hat{\Delta}$ is dense in $C(B)$; in particular, $\hat{\Delta}$ separates the points of B. For $\hat{\pi} \in \hat{\Delta}$, $\hat{\pi}(st) = \hat{\pi}(s)\hat{\pi}(t) = \hat{\pi}(t)\hat{\pi}(s) = \hat{\pi}(ts)$. Thus $st = ts$. For $\hat{\pi} \in \hat{\Delta}$, we have

$$\hat{\pi}((st)u) = \hat{\pi}(st)\hat{\pi}(u) = \hat{\pi}(s)\hat{\pi}(t)\hat{\pi}(u) = \hat{\pi}(s)\hat{\pi}(tu) = \hat{\pi}(s(tu)).$$

Thus $(st)u = s(tu)$. Thus, B is a compact abelian topological semigroup.

Let $\mu \in M(G)$. Let $\mu^{**} \in M(G)^{**}$ be defined by $\mu^{**}(F) = F(\mu)$, for $F \in M(G)^*$. Let μ^P be defined on $P \subset M(G)^*$ by restricting μ^{**} to P; that is, $\mu^P = \mu^{**} | P$. For $F \in P$, let $\hat{F}$ denote its Gelfand representation in $C(B)$. Let $\mu^B \in C(B)^* \cong M(B)$ be defined by $\mu^B(\hat{F}) = \mu^P(F)$. Let ρ denote the map from $M(G)$ to $M(B)$ defined by $\rho(\mu) = \mu^B$.

Since B is a compact abelian semigroup, $M(B)$ is a commutative Banach algebra under convolution; that is, for $\mu, \nu \in M(B)$, $\mu * \nu \in M(B)$ is defined by $\mu * \nu(f) = \int_B\int_B f(st)\, d\mu(s)\, d\nu(t)$, $f \in C(B)$. Let $\mu, \nu \in M(G)$ and $\pi \in \Delta$. Then

$$
\begin{aligned}
(\rho(\mu * \nu))(\hat{\pi}) &= (\mu * \nu)^B(\hat{\pi}) = \pi(\mu * \nu) = \pi(\mu)\pi(\nu) \\
&= \mu^B(\hat{\pi})\nu^B(\hat{\pi}) = \int_B \hat{\pi}(s)\, d\mu^B(s) \int_B \hat{\pi}(t)\, d\nu^B(t) \\
&= \int_B\int_B \hat{\pi}(s)\hat{\pi}(t)\, d\mu^B(s)\, d\nu^B(t) \\
&= \int_B\int_B \hat{\pi}(st)\, d\mu^B(s)\, d\nu^B(t) = (\mu^B * \nu^B)(\hat{\pi}).
\end{aligned}
$$

Since $Sp\hat{\Delta}$ is dense in $C(B)$, it follows that ρ preserves multiplication. Now $\| \rho(\mu) \|_{M(B)} = \| \mu^B \|_{M(B)} = \| \mu^P \|_{P^*} \leq \| \mu^{**} \|_{M(G)^{**}} = \| \mu \|_{M(G)}$. Hence $\| \rho \| \leq 1$. Also ρ is one-to-one since Δ is separating. It also follows that $M(B)$ is semisimple and that $\rho(M(G)) = M(G)^B$ is weak-$*$ dense in $M(B)$.

It remains to show that $\mu \mapsto \mu^B$ is an isometry from $M(G)$ into $M(B)$. We know that $\| \mu^B \| \leq \| \mu \|$. Now, for $\mu \in M(G)$,

$$\| \mu^B \| \geq \sup \{ | (a_1\pi_1 + \cdots + a_n\pi_n)(\mu) | : \| a_1\pi_1 + \cdots + a_n\pi_n \|_{M(G)^*} \leq 1,$$

where $\pi_i \in \Gamma\} = \sup \{ | \int_G f\, d\mu | : \| f \|_\infty \leq 1$, f a trigonometric polynomial on $G\} = \| \mu \|$.

Thus $\| \mu^B \| = \| \mu \|$. □

3.4 Definition: Let $f \in C(B)$, $f \not\equiv 0$, and $\| f \|_\infty \leq 1$, be such that $f(st) = f(s)f(t)$, $s, t \in B$. We call f a **semicharacter**. Let $\check{B}$ denote the collection of all semicharacters.

3.5 Theorem: $\hat{\Delta} = \check{B}$; *that is, the maximal ideal space of* $M(G)$ *can be identified as the set of semicharacters on the structure semigroup of* $M(G)$.

Proof: It follows, from the definition of the multiplication in B, that $\hat{\Delta} \subset \check{B}$. Let $F \in P$ be such that the Gelfand representation, $\hat{F}$, of F is in $\check{B}$. Then, for $\mu, \nu \in M(G)$,

$$\begin{aligned} F(\mu * \nu) &= \int_B \hat{F}(s)\, d(\mu * \nu)^B(s) = \int_B \hat{F}(s)\, d(\mu^B * \nu^B)(s) \\ &= \int_B\int_B \hat{F}(st)\, d\mu^B(s)\, d\nu^B(t) = \int_B\int_B \hat{F}(s)\hat{F}(t)\, d\mu^B(s)\, d\nu^B(t) \\ &= \int_B \hat{F}(s)\, d\mu^B(s) \int_B \hat{F}(t)\, d\nu^B(t) = F(\mu)F(\nu). \end{aligned}$$

Thus $F \in \Delta$ and $\hat{F} \in \hat{\Delta}$. □

3.6 Notation: Let $\hat{I} = \{f \in \hat{\Delta} : |f(s)| = 0 \text{ or } 1 \text{ for each } s \in B\}$.

3.7 Lemma: *If* $f \in \hat{\Delta}$, *then* $|f| \in \hat{\Delta}$ *and there exists a unique* $h \in \hat{I}$ *such that* $f = |f|\, h$ *where* h *has the same support as* f.

Proof: Let $h_1(s) = |f(s)|^{-1} f(s)$ if $f(s) \neq 0$, and let $h_1(s) = 0$ if $f(s) = 0$. Then h_1 is a bounded Borel function and $h_1(st) = h_1(s)h_1(t)$ for each $s, t \in B$.

Define $\pi \in \Delta$ by $\pi(\mu) = \int_B h_1(s)\, d\mu^B(s)$. Now $\pi(\mu) = \int_B \hat{\pi}(s)\, d\mu^B(s)$, and thus there is $h = \hat{\pi} \in \hat{\Delta}$ such that $h = h_1$ except on a set of μ^B-measure zero for each $\mu \in M(G)$. Since h is continuous, it follows that $h \in \hat{I}$ and $f = |f|\, h$ (recall that $\rho(M(G))$ is weak-$*$ dense in $M(B)$). □

3.8 Theorem: *If* $\mu \in M(G)$, *then the Gelfand transform,* $\tilde{\mu}$, *of* μ *attains its maximum modulus at* $\pi \in \Delta$ *such that* $\hat{\pi} \in \hat{I}$.

Proof: Since Δ is weak-$*$ compact in $M(G)^*$, $\tilde{\mu}$ attains its maximum modulus at some point $\pi \in \Delta$. By Lemma 3.7, we can write $\hat{\pi}$ as $\hat{\pi} = |\hat{\pi}| h$ where $h \in \hat{I}$. We define the complex-valued function

$$\xi(z) = \int_B |\hat{\pi}|^z(s)h(s)\,d\mu^B(s) = \mu^B(|\hat{\pi}|^z h)$$

for $\mathrm{Re}(z) > 0$, ($\mathrm{Re}(z)$ denotes the real part of z).

If $\mathrm{Re}(z_0) > 0$, then $(z_0 - z)^{-1}(x^{z_0} - x^z)$ converges uniformly for $x \in [0, 1]$ to $x^{z_0}\log(x)$ as z converges to z_0. Thus $|\hat{\pi}|^z$ has a derivative $|\hat{\pi}|^{z_0}\log(|\hat{\pi}|)$ at each point z_0 in $\{z:\mathrm{Re}(z) > 0\}$.

Thus ξ is an analytic function on $\{z: \mathrm{Re}(z) > 0\}$, [N, p. 66]. However, ξ attains its maximum modulus at $z = 1$; and thus the maximum modulus principle implies that ξ is a constant; that is, $\mu^B(|\hat{\pi}|^z h) = \mu^B(\hat{\pi})$.

Let $k(s) = \lim_n |\hat{\pi}|^n(s)$. Now $k(s) = 0$ if $|\hat{\pi}|(s) < 1$ and $k(s) = 1$ if $|\hat{\pi}|(s) = 1$. Thus k is a bounded Borel function and $k(st) = k(s)k(t)$ for all $s, t \in B$. Hence there is $\hat{\pi}_0 \in \hat{I}$ such that $\hat{\pi}_0 = k$ except on a set of μ^B-measure zero for each $\mu \in M(G)$ (as in the proof of Lemma 3.7). Now $\hat{\pi}_0 h \in \hat{I}$ and $\mu^B(\hat{\pi}) = \lim_n \mu^B(|\hat{\pi}|^n h) = \mu^B(\hat{\pi}_0 h)$ by the Lebesgue dominated convergence theorem. Hence $\tilde{\mu}$ attains its maximum modulus at a point π_1 such that $\hat{\pi}_1 = \hat{\pi}_0 h \in \hat{I}$. □

3.9 Theorem: *Let $\mu \in M^+$. For $F \in P$ with $\hat{F} \in C(B)$, let F_a denote the element of P such that $(F_a)\hat{} = |\hat{F}|$. Then $f^\mu_{F_a} = |f^\mu_F|$ in $L^\infty(\mu)$.*

Proof: Let $\eta: C(B) \to L^\infty(\mu)$ be defined by $\eta(\hat{F}) = f^\mu_F$ where $F \in P$ and $\hat{F} \in C(B)$. Now $\eta(\hat{F}\hat{H}) = \eta(\hat{F})\eta(\hat{H})$ since $f^\mu_{F\times H} = f^\mu_F f^\mu_H$; and $\eta(\overline{\hat{F}}) = \overline{\eta(\hat{F})}$ since $f^\mu_{F^\sim} = \overline{f^\mu_F}$. Also $\|\eta(\hat{F})\|_\infty \le \|\hat{F}\|_\infty$.

Now for $\hat{F} \in C(B)$, let $\phi \in P$ be such that $\phi^2 = F_a$, $(\hat{\phi})^2 = |\hat{F}|$, and $\hat{\phi} \ge 0$. Now $(\tilde{\phi})\hat{} = \overline{(\hat{\phi})} = \hat{\phi}$ so $\tilde{\phi} = \phi$. Thus $f^\mu_{\phi^\sim} = f^\mu_\phi$. Therefore

$$f^\mu_{F_a} = f^\mu_{\phi^2} = f^\mu_\phi f^\mu_\phi = f^\mu_\phi f^\mu_{\phi^\sim} = f^\mu_\phi \overline{f^\mu_\phi} \ge 0.$$

Thus $f^\mu_{F_a} \ge 0$. Consequently,

$$(f^\mu_{F_a})^2 = f^\mu_{F_a\times F_a} = f^\mu_F f^\mu_{F^\sim} = f^\mu_F \overline{f^\mu_F} = |f^\mu_F|^2.$$

Taking square roots yields that $f^\mu_{F_a} = |f^\mu_F|$. □

3.10 Corollary: *Let $z \in \mathbf{C}$ be such that* $\mathrm{Re}(z) > 0$. *Let $F_z \in P$ be such that $\hat{F}_z = |\hat{F}|^z$. Then $f^\mu_{F_z} = |f^\mu_F|^z$ in $L^\infty(\mu)$.*

Proof: By the Weierstrass approximation theorem, we may approximate $t \mapsto |t|^z$ by polynomials. Passing to the limit shows that $f^\mu_{F_z} = (f^\mu_{F_a})^z$. The previous theorem shows that $(f^\mu_{F_a})^z = |f^\mu_F|^z$. □

3.11 Corollary: *Let $\hat{\pi} \in \hat{I}$. Then $|f^\mu_\pi|$ is* 0 *or* 1 a.e. μ.

Proof: $\hat{\pi} \in \hat{I}$ implies that $|\hat{\pi}|$ is an idempotent in $C(B)$. Thus $|f^{\mu}_{\pi}|$ is an idempotent in $L^{\infty}(\mu)$ and thus is 0 or 1 a.e. μ. □

3.12 Definition: The **Shilov boundary,** ∂M, of $M(G)$ is the unique minimal closed subset of Δ on which each $\tilde{\mu}$ achieves its maximum modulus. Theorem 3.8 implies:

3.13 Theorem: Let $I = \{\pi \in \Delta : \hat{\pi} \in \hat{I}\}$. Then $\bar{I} \supset \partial M$.

3.14 Remark: In the next chapter, it is shown that $\bar{I}$ is a proper subset of Δ when G is nondiscrete.

4. Generalized Characters

4.1 Definition: A **generalized function** is a collection of functions $\{g^{\mu} : \mu \in M^{+}\}$ such that $g^{\mu} \in L^{\infty}(\mu)$, such that if $\nu \ll \mu$, then $g^{\nu} = g^{\mu}$ a.e. $\nu(\mu, \nu \in M^{+})$, and such that $\sup\{\| g^{\mu} \|_{\infty} : \mu \in M^{+}\} < \infty$. The collection of such is denoted by $\mathscr{F}$. For $\{g^{\mu}\} \in \mathscr{F}$, let $\| \{g^{\mu}\} \| = \sup\{\| g^{\mu} \|_{\infty} : \mu \in M^{+}\}$.

If, in addition, $\| \{g^{\mu}\} \| = 1$ and $g^{\mu}(x + y) = g^{\mu}(x)g^{\mu}(y)$ a.e. $\mu \times \mu$ on $G \times G$ for all $\mu \in M^{+}$ with $\mu * \mu \ll \mu$, then $\{g^{\mu}\}$ is called a **generalized character**. The collection of such is denoted by $\mathscr{C}$.

4.2 Theorem: *The map $\eta : M(G)^{*} \to \mathscr{F}$ defined by $\eta(F) = \{f^{\mu}_{F} : \mu \in M^{+}\}$ is isometric and *-isomorphic.*

Proof: Let $F \in M(G)^{*}$. For $\nu \ll \mu(\nu, \mu \in M^{+})$ and $\lambda \ll \nu$, we have that

$$\int_G f^{\nu}_{F}\, d\lambda = F(\lambda) = \int_G f^{\mu}_{F}\, d\lambda.$$

Thus $f^{\nu}_{F} = f^{\mu}_{F}$ a.e. ν. Also $\| \{f^{\mu}_{F} : \mu \in M^{+}\} \| = \| F \| < \infty$. So η maps into $\mathscr{F}$.

Let $\{g^{\mu} : \mu \in M^{+}\} \in \mathscr{F}$. Define $F(\nu) = \int_G g^{\nu}\, d\nu$, $\nu \in M^{+}$. Now F is linear on M^{+} since for $a \geq 0$ and $\nu, \lambda \in M^{+}$ we have

$$\begin{aligned} F(a\nu + \lambda) &= \int_G g^{(a\nu + \lambda)}\, d(a\nu + \lambda) \\ &= a\int_G g^{(a\nu + \lambda)}\, d\nu + \int_G g^{(a\nu + \lambda)}\, d\lambda \\ &= a\int_G g^{\nu}\, d\nu + \int_G g^{\lambda}\, d\lambda \\ &= aF(\nu) + F(\lambda). \end{aligned}$$

Now extend linearly to all of $M(G)$. Now F is bounded since

$$\| F \| = \| \{g^{\mu} : \mu \in M^{+}\} \| < \infty.$$

The remainder of the theorem follows from Lemmas 2.6 and 2.8. Note that for $\nu \in M(G)$,

$$F(\nu) = F(\nu_1) - F(\nu_2) + iF(\nu_3) - iF(\nu_4)$$

(where $\nu = \nu_1 - \nu_2 + i\nu_3 - i\nu_4$ and each $\nu_i \geq 0$)

$$= \int_G g^{\nu_1}\,d\nu_1 - \int_G g^{\nu_2}\,d\nu_2 + i\int_G g^{\nu_3}\,d\nu_3 - i\int_G g^{\nu_4}\,d\nu_4$$
$$= \int_G g^{|\nu|}\,d\nu_1 - \int_G g^{|\nu|}\,d\nu_2 + i\int_G g^{|\nu|}\,d\nu_3 - i\int_G g^{|\nu|}\,d\nu_4$$

(where $|\nu|$ denotes the total variation of ν)

$$= \int_G g^{|\nu|}\,d\nu.$$

Thus $F(\nu) = \int_G g^{|\nu|}\,d\nu$. □

4.3 Theorem: *The map $\eta:\Delta \to \mathscr{C}$ defined by $\eta(\pi) = \{f_\pi^\mu : \mu \in M^+\}$ is one-to-one and onto.*

Proof: Let $\{g^\mu : \mu \in M^+\}$ be a generalized character. Define

$$\pi(\nu) = \int_G g^{|\nu|}\,d\nu,$$

where $|\nu|$ denotes the total variation of ν. As in Theorem 4.2, $\pi \in M(G)^*$. We show that π is multiplicative.

Let $\mu, \nu \in M(G)$. Let $\lambda = \exp(|\mu| + |\nu|)$. Now $\mu, \nu \ll \lambda$. Since $\lambda * \lambda \ll \lambda$, $g^\lambda(x + y) = g^\lambda(x)g^\lambda(y)$ a.e. $\lambda \times \lambda$. Thus

$$\pi(\mu * \nu) = \int_G g^{|\mu*\nu|}\,d\mu * \nu = \int_G g^{\lambda*\lambda}\,d\mu * \nu$$
$$= \int_G g^\lambda\,d\mu * \nu = \int_G\int_G g^\lambda(x + y)\,d\mu(x)\,d\nu(y)$$
$$= \int_G\int_G g^\lambda(x)g^\lambda(y)\,d\mu(x)\,d\nu(y)$$
$$= \int_G g^\lambda(x)\,d\mu(x)\int_G g^\lambda(y)\,d\nu(y)$$
$$= \int_G g^{|\mu|}\,d\mu\int_G g^{|\nu|}\,d\nu = \pi(\mu)\pi(\nu).$$

Let $\pi \in \Delta$ and $\mu \in M^+$ where $\mu * \mu \ll \mu$. We must show that $f_\pi^\mu(x + y) = f_\pi^\mu(x)f_\pi^\mu(y)$ a.e. $\mu \times \mu$. Now since π is multiplicative, for $\mu', \mu'' \ll \mu$ (so $\mu' * \mu'' \ll \mu * \mu \ll \mu$) we have that

$$\int_G\int_G f_\pi^\mu(x + y)\,d\mu'(x)\,d\mu''(y) = \int_G f_\pi^{|\mu'*\mu''|}(x)\,d(\mu' * \mu'')(x)$$
$$= \pi(\mu' * \mu'') = \pi(\mu')\pi(\mu'')$$
$$= \int_G f_\pi^{|\mu'|}(x)\,d\mu'(x)\int_G f_\pi^{|\mu''|}(y)\,d\mu''(y)$$
$$= \int_G f_\pi^\mu(x)\,d\mu'(x)\int_G f_\pi^\mu(y)\,d\mu''(y)$$
$$= \int_G\int_G f_\pi^\mu(x)f_\pi^\mu(y)\,d\mu'(x)\,d\mu''(y).$$

Thus $f_\pi^\mu(x + y) = f_\pi^\mu(x)f_\pi^\mu(y)$ a.e. $\mu \times \mu$. □

5. Historical Notes

5.1: The structure of the maximal ideal space of $M(G)$ was first studied by Šreider [1]. He represented Δ as a space of generalized characters on G. We have modified slightly the definition of a generalized character.

J. Taylor [1] showed that Δ could be represented as the collection of all continuous semicharacters on some compact abelian topological semigroup. For this result, we have essentially followed the proof from Ramirez [4]. We avoided here the introduction of the Arens multiplication by following the observation due to I. Glicksberg that the Arens multiplication in our case is computable (see Lemma 2.6 and Theorem 2.9). Taylor used the L-space theory developed by Kakutani. Our results, although not as general as Taylor's, are less technical.

The study of $\hat{I}$ (Theorem 3.8) is due to J. Taylor, [1, p. 162].

The measure algebra on locally compact abelian semigroups (separately continuous) has been studied, using the Arens product, by Rennison [1].

The Choquet boundary of $M(G)$ has been studied by Miller [1].

CHAPTER 2

THE SHILOV BOUNDARY AND THE SYMMETRIC MAXIMAL IDEALS IN M(G)

1. Introduction

1.1: In this chapter, G will be a **nondiscrete** LCA group. The notation established in Chapter 1 will be carried over. Now $M(G)$ has a natural involution defined by $\mu^*(E) = \overline{\mu(-E)}$, E Borel in G. Since $M(G)$ is an asymmetric algebra, the symmetric maximal ideals, $\mathscr{S}$, of $M(G)$ are properly contained in the maximal ideal space, Δ, of $M(G)$. The main result of this chapter [Theorem 4.4] is the fact that there are symmetric maximal ideals not contained in the Shilov boundary, ∂M, of $M(G)$. In particular, $\partial M \neq \Delta$ and $\bar{\Gamma} \neq \mathscr{S}$ ($\bar{\Gamma}$ denotes the closure of Γ in Δ). Thus

$$\bar{\Gamma} \subsetneqq \partial M \subsetneqq \Delta,$$

$$\bar{\Gamma} \subsetneqq \mathscr{S} \subsetneqq \Delta,$$

$$\mathscr{S} \not\subset \partial M, \text{ and}$$

$$\partial M \not\subset \mathscr{S}.$$

Furthermore, $\Delta \neq \mathscr{S} \cup \partial M$ [Remark 4.6].

To derive the main result, one shows the existence of $\mu \geq 0$ satisfying three properties (A), (B), and (C) [Theorem 2.4]. Since these conditions are not easily computed, one utilizes three other conditions (A′), (B′), and (C′) [Lemmas 2.7, 2.8, 2.10]. One then shows the existence of the required μ for an infinite product of cyclic groups [Theorem 3.1], for the p-adic integers [Theorem 3.3], for the reals modulo 1 [Theorem 3.4], and for the reals [Theorem 3.5]. The general case is then deduced [Theorem 4.4].

The method is to consider for a fixed $\mu \geq 0$ the sets

$$D(\overline{\Gamma}) = \{ |a| : 0 \neq a = f_\pi^\mu \text{ in } L^\infty(\mu), \pi \in \overline{\Gamma}\},$$
$$D(\partial M) = \{ |a| : 0 \neq a = f_\pi^\mu \text{ in } L^\infty(\mu), \pi \in \partial M\},$$

and

$$D(\mathscr{S}) = \{ |a| : 0 \neq a = f_\pi^\mu \text{ in } L^\infty(\mu), \pi \in \mathscr{S}\}.$$

Property (A) will yield that $(0, 1) \subset D(\mathscr{S})$. Property (B) will yield that $D(\overline{\Gamma}) = D(\partial M)$. Property (C) will yield that $D(\overline{\Gamma}) \subsetneqq (0, 1)$. Thus for a particular μ, we will have

$$D(\overline{\Gamma}) = D(\partial M) \subsetneqq (0, 1) \subset D(\mathscr{S}).$$

This gives the desired result that $\mathscr{S} \not\subset \partial M$.

1.2 Definition: Let $\mathscr{S}$ denote the **symmetric maximal ideals** in $M(G)$; that is $\mathscr{S} = \{\pi \in \Delta : \tilde{\mu}^*(\pi) = \overline{\tilde{\mu}(\pi)} \text{ for all } \mu \in M(G)\}$.

1.3 Remarks:

(i) The set $\mathscr{S}$ is a closed subset of Δ.

(ii) On $\mathscr{S}$, the Gelfand transforms of $M(G)$, $M(G)\tilde{}$, are closed under conjugation, and thus by the Stone-Weierstrass theorem, $M(G)\tilde{} \,|\, \mathscr{S}$ is sup-norm dense in $C(\mathscr{S})$.

(iii) Since the Fourier-Stieltjes transforms satisfy

$$\hat{\mu}^* = \overline{\hat{\mu}}, \Gamma \subset \overline{\Gamma} \subset \mathscr{S}.$$

(iv) For G a nondiscrete LCA group, $M(G)$ is an asymmetric algebra [R, p. 107] and thus $\mathscr{S} \neq \Delta$. We now show that $\partial M \not\subset \mathscr{S}$.

1.4 Theorem: *Let G be a nondiscrete* LCA *group. Then not all maximal ideals in the Shilov boundary are symmetric; that is, $\partial M \not\subset \mathscr{S}$. In particular, $\overline{\Gamma} \subsetneqq \partial M$.*

Proof: Suppose that all the maximal ideals in the Shilov boundary are symmetric. We will derive a contradiction by showing that this implies that $\partial M = \Delta$ and so $M(G)$ would be symmetric. Let $\pi \in \Delta$. The map $\tilde{\mu} \,|\, \partial M \mapsto \tilde{\mu}(\pi)$ is a multiplicative linear functional on a dense subalgebra of $C(\partial M)$ which extends uniquely to point evaluation at some point in ∂M. Thus $\pi \in \partial M$. □

2. Preparatory Results

2.1 Notation: Let $\pi \in \Delta$ and $\mu \in M^+$. Define

$$f_\pi^\mu \in L^1(\mu)^* = L^\infty(\mu) \text{ by } \pi(\nu) = \int_G f_\pi^\mu \, d\nu, \; \nu \in L^1(\mu), \text{ [see 1.2.5]}.$$

For $\gamma \in \Gamma \subset \Delta, f^\mu_\gamma = \gamma$ in $L^\infty(\mu)$ for any $\mu \geq 0$.

For $a \in \mathbf{C}$, $\mathbf{a}$ will denote the constant function of value a. Let

$$D_\mu = \{a \in \mathbf{C} : 0 \neq \mathbf{a} \in \overline{\Gamma^\sigma}\},$$

the closure in the $\sigma(L^\infty(\mu), L^1(\mu))$ topology; that is, the weak-$*$ topology in $L^\infty(\mu)$ from $L^1(\mu)$.

We now characterize the symmetric maximal ideals from their representation as generalized characters, [see 1.4.3].

2.2 Lemma: *The generalized character $\{f^\mu_\pi : \mu \in M^+\}$ corresponds to a symmetric maximal ideal if and only if for each $\sigma \in M^+$ with $\sigma = \sigma^*$ we have that*

$$f^\sigma_\pi(t) = \overline{f^\sigma_\pi(-t)}$$

in $L^\infty(\sigma)$.

Proof: First, suppose $\{f^\mu_\pi : \mu \geq 0\}$ corresponds to a symmetric maximal ideal. Now for $\sigma \in M(G)$ with $\sigma \geq 0$ and $\sigma = \sigma^*$, $d\sigma(t) = d\sigma(-t)$. Thus for $g \in L^1(\sigma)$,

$$\int_G f^\sigma_\pi g \, d\sigma = \pi(g \, d\sigma) = \overline{\pi(g^* \, d\sigma)} = \overline{\int_G f^\sigma_\pi g^* \, d\sigma} = \int_G \overline{f^\sigma_\pi}(-t) g(t) \, d\sigma(t).$$

Now suppose $f^\sigma_\pi(t) = \overline{f^\sigma_\pi(-t)}$ in $L^\infty(\sigma)$ when $\sigma \geq 0$ with $\sigma = \sigma^*$. Let $\mu \in M(G)$. We consider $\nu = |\mu| + |\mu^*|$. Then $\nu \geq 0$ and $\nu = \nu^*$. Thus for $g \in L^1(\nu)$ where $g \, d\nu = \mu$, we have

$$\begin{aligned} \pi(\mu) = \pi(g \, d\nu) = \int_G f^\nu_\pi g \, d\nu &= \overline{\int_G \overline{f^\nu_\pi(-t)} g(-t) \, d\nu(-t)} \\ &= \overline{\int_G f^\nu_\pi(t) g^*(t) \, d\nu(t)} = \overline{\pi(g^* \, d\nu)} \\ &= \overline{\pi((g \, d\nu)^*)} = \overline{\pi(\mu^*)}. \end{aligned}$$

Thus π is a symmetric maximal ideal. □

2.3 Lemma: *Let $\mu \in M(G)$ with $\mu \geq 0$.*

(i) *$D_\mu = \{a \in \mathbf{C} : 0 \neq \mathbf{a} = f^\mu_\pi$ in $L^\infty(\mu), \pi \in \overline{\Gamma}\}$.*

(ii) *If (A) $\equiv$ (there is $a \in D_\mu$ such that $0 < |a| < 1$), then*

$$\{|a| : 0 \neq \mathbf{a} = f^\mu_\pi \text{ in } L^\infty(\mu) \text{ with } \pi \in \mathcal{S}\} \supset \quad (0, 1).$$

(iii) *If (B) $\equiv$ (for every $\pi \in \Delta$, there is $a \in \mathbf{C}$ and $\gamma \in \Gamma$ such that $f^\mu_\pi = a\gamma$ in $L^\infty(\mu)$), then*

$$\begin{aligned} &\{|a| : 0 \neq \mathbf{a} = f^\mu_\pi \text{ in } L^\infty(\mu) \text{ with } \pi \in \partial M\} \\ &\qquad = \{|a| : 0 \neq \mathbf{a} = f^\mu_\pi \text{ in } L^\infty(\mu) \text{ with } \pi \in \overline{\Gamma}\}. \end{aligned}$$

Proof: Let $\pi \in \overline{\Gamma}$ with $f^\mu_\pi = \mathbf{a} \neq 0$. Then $a \in D_\mu$ since convergence in Δ is stronger than convergence in $\sigma(L^\infty(\mu), L^1(\mu))$. Now let $a \in D_\mu$. Let $\{\gamma_\alpha\} \subset \Gamma$

be such that $\gamma_\alpha \xrightarrow{\alpha} \mathbf{a}$ in $\sigma(L^\infty(\mu), L^1(\mu))$. Since $\Gamma \subset \Delta$ and Δ is compact, there is a convergent subnet $\{\gamma_\beta\}$ of $\{\gamma_\alpha\}$ such that $\gamma_\beta \xrightarrow{\beta} \pi$ in $\overline{\Gamma} \subset \Delta$. Now

$$f_\pi^\mu = \mathbf{a} \text{ in } L^\infty(\mu)$$

since for $v \in L^1(\mu)$,

$$\begin{aligned}\textstyle\int_G f_\pi^\mu \, dv &= \pi(v) = \lim_\beta \hat{v}(\gamma_\beta)\\ &= \lim_\beta \textstyle\int_G \gamma_\beta \, dv = \int_G \mathbf{a} \, dv.\end{aligned}$$

Thus (i) holds.

Let $\pi \in \overline{\Gamma}$ with $f_\pi^\mu = \mathbf{a} \neq 0$ in $L^\infty(\mu)$ where $0 < |a| < 1$. For $0 < x < \infty$, let $\pi_x \in \Delta$ be such that $\hat{\pi}_x = |\hat{\pi}|^x$ in $C(B)$ and thus $f_{\pi_x}^\mu = |f_\pi^\mu|^x = |\mathbf{a}|^x$ [see 1.3.10]. By Lemma 2.2, π_x is symmetric. As x ranges over $(0, \infty)$, $|a|^x$ ranges over (0, 1). Thus (ii) holds.

Let $\pi \in \partial M$ with $f_\pi^\mu = \mathbf{b} \neq 0$. We show that $|b| = |a|$ for some $a \in D_\mu$. Now there is a net $\{\pi_\alpha\} \subset I$ such that $\pi_\alpha \xrightarrow{\alpha} \pi$ in Δ since $\overline{I} \supset \partial M$ [see 1.3.13]. Now $f_{\pi_\alpha}^\mu \xrightarrow{\alpha} \mathbf{b}$ in $\sigma(L^\infty(\mu), L^1(\mu))$.

Write $f_{\pi_\alpha}^\mu = a_\alpha \gamma_\alpha$, $a_\alpha \in \mathbf{C}$, $\gamma_\alpha \in \Gamma$ by condition (B). Now $\pi_\alpha \in I$ implies that $|a_\alpha| = 0$ or 1 since

$$|f_{\pi_\alpha}^\mu| = |a_\alpha \gamma_\alpha| = |a_\alpha| \text{ is 0 or 1 a.e. } \mu$$

[see 1.3.11]. By transferring to a subnet if necessary, we can assume that $c = \lim_\alpha a_\alpha$ exists and where $|c| = 0$ or 1. Now $c\gamma_\alpha \xrightarrow{\alpha} \mathbf{b} \neq 0$ in $\sigma(L^\infty(\mu), L^1(\mu))$ and so $|c| = 1$; and letting $a = bc^{-1}$, we have that $a \in D_\mu$ and $|a| = |b|$. This suffices for (iii) since $\overline{\Gamma} \subset \partial M$. □

2.4 Theorem: *Suppose there is $\mu \in M^+$ such that*

(A) ≡ (*there is $a \in D_\mu$ with $0 < |a| < 1$*),

(B) ≡ (*for every $\pi \in \Delta$, there is $a \in \mathbf{C}$ and $\gamma \in \Gamma$ such that*

$$f_\pi^\mu = a\gamma \text{ in } L^\infty(\mu)), \text{ and}$$

(C) ≡ ($\{|a| : a \in D_\mu\}$ *does not contain* (0, 1)).

Then there is $\pi \in \mathscr{S}$ such that $\pi \notin \partial M$; that is, $\mathscr{S} \not\subset \partial M$.

Proof: $\{|a| : 0 \neq \mathbf{a} = f_\pi^\mu, \pi \in \partial M\} = \{|a| : a \in D_\mu\}$ and $\{|a| : a \in D_\mu\}$ does not contain (0, 1). But

$$\{|a| : 0 \neq \mathbf{a} = f_\pi^\mu, \pi \in \mathscr{S}\}$$

does contain (0, 1). So $\mathscr{S} \not\subset \partial M$. □

2.5 Lemma: *Let $\{V_i\}_{i=0}^{\infty}$ be a basic set of compact neighborhoods of 0 such that $V_{i+1} + V_{i+1} \subset V_i$. Let μ_i be a probability measure with support in V_i. Then*

$$\mathop{\bigstar}_{i=1}^{\infty} \mu_i$$

converges in the weak-$*$ topology to a probability measure, μ, whose support is contained in $V_1 + V_1 \subset V_0$.

Proof: Let $\lambda_n = \mu_1 * \mu_2 * \cdots * \mu_n$. For $m > n$, let $\lambda_{m,n} = \mu_{n+1} * \cdots * \mu_m$. Now spt $(\lambda_{m,n}^*)$ is contained in $V_{n+1} + V_{n+1}$, and therefore for f uniformly continuous,

$$\lambda_{m,n}^* * f \xrightarrow{m,n} f$$

uniformly. Now for $f \in C_0(G)$, $\{\lambda_n(f)\}$ is a Cauchy sequence since

$$|\lambda_n(f) - \lambda_m(f)| = |\lambda_n(f) - \lambda_n(\lambda_{m,n}^* * f)|$$
$$\leq \|f - \lambda_{m,n}^* * f\|_\infty \xrightarrow{m,n} 0.$$

Define $\mu(f) = \lim_n \lambda_n(f)$. Clearly μ is a bounded linear functional on $C_0(G)$ with spt $(\mu) \subset V_1 + V_1$. □

2.6 Notation: Suppose we have $V_0, V_1, \ldots,$ a basic system of compact neighborhoods of 0 in G with $V_i \supset V_{i+1} + V_{i+1}$; $m_1, m_2, \ldots,$ positive integers and x_{ij}, $i = 1, 2, \ldots, j = 0, 1, \ldots, m_i - 1$ such that

$$x_{i0} = 0,\ x_{ij} \in V_{i-1},$$

and V_{i-1} is equal to the union of the disjoint sets $x_{ij} + V_i, j = 0, \ldots, m_i - 1$. Let

$$X_i = \{x_{ij} : j = 0, \ldots, m_i - 1\}, \qquad Y_i = X_1 + X_2 + \cdots + X_i,$$
$$M_i = \frac{2}{m_i(m_i + 1)},\ \delta_i = M_i \sum_{j=0}^{m_i - 1} (m_i - j)\delta_{x_{ij}}$$

where δ_x denotes the unit point mass at x,

$$\mu = \mathop{\bigstar}_{i=1}^{\infty} \delta_i, \text{ and } \mu_n = \mathop{\bigstar}_{i=n+1}^{\infty} \delta_i.$$

Lemma 2.5 implies that μ and μ_n exist. Also μ_n is supported in V_n since $X_{i+1} + X_{i+2} + \cdots + X_{i+j} \subset V_i$. If ν is a probability measure on V_i, then for $y \in Y_i$,

$$\delta_1 * \cdots * \delta_i(\{y\}) = \delta_1 * \cdots * \delta_i * \nu(y + V_i).$$

Thus

$$\mu = \left(\sum_{y \in Y_i} \mu(y + V_i)\delta_y \right) * \mu_i.$$

Let S be the countable subgroup of G generated by the x_{ij}. Let v be a bounded measure obtained by placing positive point masses at all the points of S. Let $\sigma = \mu + v$. One introduces v since f_{π}^{μ} is not well defined at the points x_{ij}.

For $y \in Y_i$, let $c(i, y)$ denote the characteristic function of $y + V_i$ and let C_i denote the linear span of the $c(i, y)$. Since $\bigcap_{i=0}^{\infty} V_i = \{0\}$ and since V_{i-1} is the union of the disjoint sets $x_{ij} + V_i$, each continuous function on V_0 is the uniform limit of functions from $\bigcup_{i=0}^{\infty} C_i$. Since $\operatorname{spt}(\mu) \subset V_0$, $\bigcup_{i=0}^{\infty} C_i$ is dense in $L^1(\mu)$.

2.7 Lemma: *With the previous notation, the following holds:*

$$(C') \equiv (D_u \subset (\hat{\mu}(\Gamma))^{-}).$$

Proof: If $a \in D_\mu$, then there is a net $\{\gamma_\alpha\} \subset \Gamma$ with $\gamma_\alpha \xrightarrow{\alpha} \mathbf{a}$ in $\sigma(L^\infty(\mu), L^1(\mu))$. In particular,

$$\hat{\mu}(\gamma_\alpha) = \int_G \gamma_\alpha \, d\mu \xrightarrow{\alpha} \int_G \mathbf{a} \, d\mu = a.$$

So $a \in (\hat{\mu}(\Gamma))^{-}$. □

2.8 Lemma: *With the previous notation, if* (A') $\equiv$ (*suppose there is a net* $\{\gamma_\alpha\} \subset \Gamma$ *such that* $\hat{\mu}(\gamma_\alpha) \xrightarrow{\alpha} a$ *and also* $\gamma_\alpha(s) \xrightarrow{\alpha} 1$ *for each* $s \in S$), *then*

$$a \in D_\mu.$$

Proof: Since $\| \hat{\mu} \|_\infty \leq \| \mu \| = 1$, $| a | \leq 1$. Since $\bigcup_{i=0}^{\infty} C_i$ is dense in $L^1(\mu)$ and since $\| \gamma_\alpha \|_\infty \leq 1$, we need only show that

$$\int_G c(i, y)\gamma_\alpha \, d\mu \xrightarrow{\alpha} \int_G c(i, y)\mathbf{a} \, d\mu = \mu(y + V_i)a.$$

Now

$$\int_G \gamma_\alpha c(i, y) \, d\mu = \int_G \gamma_\alpha c(i, y) \, d\left(\sum_{z \in Y_i} \mu(z + V_i)\delta_z \right) * \mu_i$$

$$= \mu(y + V_i)\gamma_\alpha(y)\int_G \gamma_\alpha \, d\mu_i$$

since $\operatorname{spt}(\mu_i) \subset V_i$. Note that

$$\int_G \gamma_\alpha \, d\mu = \left(\int_G \gamma_\alpha \, d\mu_i\right)\left(\sum_{z \in Y_i} \gamma_\alpha(z)\mu(z + V_i) \right).$$

Since $\int_G \gamma_\alpha \, d\mu \xrightarrow{\alpha} a$, and since

$$\sum_{z \in Y_i} \gamma_\alpha(z)\mu(z + V_i) \xrightarrow{\alpha} 1,$$

it follows that $\int_G \gamma_\alpha \, d\mu_i \xrightarrow{\alpha} a$. Hence

$$\int_G \gamma_\alpha c(i, y) \, d\mu \xrightarrow{\alpha} \mu(y + V_i)a. \quad \square$$

2.9 Lemma: *f_π^σ is a character on S.*

Proof: Let $s, t \in S$. Now

$$\begin{aligned} f_\pi^\sigma(s+t) &= \int_G f_\pi^\sigma \, d\delta_{s+t} = \pi(\delta_{s+t}) \\ &= \pi(\delta_s * \delta_t) = \pi(\delta_s)\pi(\delta_t) = \int_G f_\pi^\sigma \, d\delta_s \int_G f_\pi^\sigma \, d\delta_t \\ &= f_\pi^\sigma(s) f_\pi^\sigma(t). \quad \square \end{aligned}$$

2.10 Lemma: *Let $\pi \in \Delta$. If* (B′) $\equiv (M_n \Sigma_j (m_n - j) \,|\, b_n f_\pi^\sigma(x_{nj}) - b_{n-1} \,| \xrightarrow{n} 0$ *where $\{b_n\}$ is a sequence of nonzero complex numbers with constant absolute value implies there is $\gamma \in \Gamma$ such that on S, $f_\pi^\sigma = \gamma$), then* (B) (see 2.4).

Proof: Let $\pi \in \Delta$. Now $f_\pi^\sigma = f_\pi^\mu$ a.e. μ since $\mu \ll \sigma$. Thus $f_\pi^\sigma \in L^\infty(\mu) \subset L^1(\mu)$. We may assume that $f_\pi^\sigma \neq 0$.

Since $\bigcup_{i=0}^{\infty} C_i$ is dense in $L^1(\mu)$, there are $f_i \in C_i$ such that

$$\| f_i - f_\pi^\sigma \|_1 \xrightarrow{i} 0 \text{ in } L^1(\mu)$$

and where f_i is chosen such that the norm is minimal. This we can do since C_i is finite dimensional.

Write $f_i = \Sigma_{y \in Y_i} \; a(i, y) \; c \; (i, y)$ where $a(i, y)$ are complex numbers. Now

$$\begin{aligned} \| f_i - f_\pi^\sigma \|_1 &= \sum_{y \in Y_i} \int_{y+V_i} | \mathbf{a}(i, y) - f_\pi^\sigma(z) | \, d\mu(z) \\ &= \sum_{y \in Y_i} \int_{y+V_i} | \mathbf{a}(i, y) - f_\pi^\sigma(z) | \, d(\mu(y + V_i)\delta_y * \mu_i)(z) \\ &= \sum_{y \in Y_i} \mu(y + V_i) \int_{y+V_i} | \mathbf{a}(i, y) - f_\pi^\sigma(z) | \, d\mu_i(z - y) \\ &= \sum_{y \in Y_i} \mu(y + V_i) \int_{V_i} | \mathbf{a}(i, y) - f_\pi^\sigma(y + t) | \, d\mu_i(t) \\ &= \sum_{y \in Y_i} \mu(y + V_i) \int_{V_i} | a(i, y) f_\pi^\sigma(y)^{-1} - f_\pi^\sigma(t) | \, d\mu_i(t) \end{aligned}$$

since f_π^σ is a character on $S \supset Y_i$. Since the $a(i, y)$ are chosen to make $\| f_i - f_\pi^\sigma \|_1$ minimal,

$$\int_{V_i} | a(i, y) f_\pi^\sigma(y)^{-1} - f_\pi^\sigma(t) | \, d\mu_i(t)$$

have the same value for each $y \in Y_i$. Hence $a(i, y) f_\pi^\sigma(y)^{-1}$ have a common value. Thus $a(i, y) = a_i f_\pi^\sigma(y)$ and so $f_i = a_i \Sigma_{y \in Y_i} f_\pi^\sigma(y) c(i, y)$.

Since $\| f_m - f_n \|_1 \geq | \, |a_m| - |a_n| \, |$, the $\lim_m |a_m|$ exists and we denote it by A. If $A = 0$, then $f_\pi^\sigma = 0$ in $L^\infty(\mu)$. Thus $A \neq 0$. We may assume

$|a_i| \neq 0$ for all i. Let $c_i = Aa_i/|a_i|$ and $g_i = c_i \Sigma_{y \in Y_i} f^\sigma_\pi(y)c(i, y)$. Thus each $|c_i| = A$. Now

$$\begin{aligned} \| g_n - g_{n-1} \|_1 &\leq \| g_n - f_n \|_1 + \| f_n - f_{n-1} \|_1 + \| f_{n-1} - g_{n-1} \|_1 \\ &\leq |c_n - a_n| + \| f_n - f_{n-1} \|_1 + |c_{n-1} - a_{n-1}| \xrightarrow{n} 0. \end{aligned}$$

Also

$$\begin{aligned} \| g_n - g_{n-1} \|_1 &= \sum_{y \in Y_{n-1}} \sum_{x \in X_n} |c_n f^\sigma_\pi(x + y) - c_{n-1} f^\sigma_\pi(y)| \, \mu(x + y + V_n) \\ &= \sum_{x \in X_n} \sum_{y \in Y_{n-1}} |c_n f^\sigma_\pi(x) - c_{n-1}| \, \mu(x + y + V_n) \\ &= \sum_{x \in X_n} \sum_{y \in Y_{n-1}} |c_n f^\sigma_\pi(x) - c_{n-1}| \, \delta_1 * \cdots * \delta_n(\{x + y\}) \\ &= \sum_j |c_n f^\sigma_\pi(x_{nj}) - c_{n-1}| \sum_{y \in Y_{n-1}} \delta_1 * \cdots * \delta_n(\{y + x_{nj}\}) \\ &= \sum_j |c_n f^\sigma_\pi(x_{nj}) - c_{n-1}| \, \delta_n(\{x_{nj}\}) \\ &= M_n \sum_j (m_n - j) \, |c_n f^\sigma_\pi(x_{nj}) - c_{n-1}|. \end{aligned}$$

By (B′) there is $\gamma \in \Gamma$ with $\gamma = f^\sigma_\pi$ on S. Since γ is continuous, $c_i^{-1} g_i = \Sigma_{y \in Y_i} f^\sigma_\pi(y)c(i, y) = \Sigma_{y \in Y_i} \gamma(y)c(i, y) \xrightarrow{i} \gamma$ uniformly on V_0. Thus $c_i^{-1} g_i \xrightarrow{i} \gamma$ in $L^1(\mu)$. Thus the limit in $L^1(\mu)$ of $c_i \gamma$ is the same as the limit of g_i, which is f^σ_π. Thus $c = \lim_i c_i$ exists and $c\gamma = f^\sigma_\pi = f^\mu_\pi$ in $L^\infty(\mu)$.□

3. Particular Cases

3.1 Theorem: *Let G be a countable product of finite cyclic groups. Then $\mathscr{S} \not\subset \partial M$.*

Proof: Let $Z(n)$ be the group of integers modulo n. There are $m_i \geq 2$ such that $G = Z(m_1) \times Z(m_2) \times \cdots$. Let $V_i = \{g \in G : g_j = 0 \text{ for } j \leq i\}$ and let $x_{ij} = (0, \ldots, 0, j, 0, \ldots)$ where $0 \leq j \leq m_i - 1$ appears in the ith coordinate.

It will suffice to verify (A), (B), and (C). We will utilize (A′), (B′), and (C′). First we verify (B′). Let $\pi \in \Delta$. Now X_n is isomorphic to $Z(m_n)$. Since f^σ_π is a character on S, $f^\sigma_\pi(x_{nj}) = (w_n)^j$ where w_n is some m_nth root of unity. Let $K_n = M_n \Sigma_j (m_n - j) \, |b_n(w_n)^j - b_{n-1}|$ with $|b_n|$ constant and nonzero. Assume $K_n \xrightarrow{n} 0$. For (B′), it will suffice to show that $w_n = 1$ for large n. If $w_n \neq 1$ for arbitrarily large n, then $|b_n(w_n)^j - b_{n-1}| \geq \frac{1}{2}|b_n|$ for at least half of the j's (recall that $|b_n|$ is constant). Now the sum of at least half of the terms in the sequence $\{M_n, 2M_n, \ldots, m_nM_n\}$ is at least

$$\frac{2}{m_n(m_n+1)}\left(1+\left[\frac{m_n+1}{2}\right]\right)\left(\left[\frac{m_n+1}{2}\right]\Big/2\right) \geq \frac{1}{4}$$

where $[x]$ denotes the greatest integer $\leq x$. Thus $K_n \geq \frac{1}{8}|b_n|$ which is the required contradiction. Thus (B) holds.

The values of $\hat{\mu}$ are 1 or a finite product of numbers of the form

$$\alpha = \frac{2}{m(m+1)} \sum_{j=0}^{m-1} (m-j)w^j$$

where w is an mth root of unity other than 1. Now

$$m = m - w(1 + w + \cdots + w^{m-1}) = (1-w) \sum_{j=0}^{m} (m-j)w^j.$$

Thus

$$\alpha = \frac{2}{m+1}\frac{1}{1-w}.$$

Also

$$|1-w| \geq 2\left|\sin\left(\frac{\pi}{m}\right)\right|.$$

Hence

$$|\alpha| \leq \frac{1}{m+1}\frac{1}{\sin(\pi/m)} \xrightarrow{m} \frac{1}{\pi}.$$

It follows that the values of $|\hat{\mu}|$ are bounded away from 1 and hence are not dense in (0, 1). Thus (C) follows from (C′). One should note that the particular choice of weights for μ are needed in applying (C′).

Let

$$\alpha_j = \frac{2}{(m_j+1)}\frac{1}{1-\exp(2\pi i/m_j)}.$$

From the above paragraph we see that $\{\alpha_j\}$ has a cluster point a with $0 < |a| < 1$. Choose a subsequence j_k such that $\alpha_{j_k} \xrightarrow{k} a$. Let $\gamma_k \in \hat{G}$ be defined by

$$\gamma_k(g) = \exp\left(\frac{2\pi i}{m_{j_k}} g_{j_k}\right).$$

Let $s \in S$. For large k, we have $\gamma_k(s) = 1$. Thus $\gamma_k \xrightarrow{k} 1$ pointwise on S and $\hat{\mu}(\gamma_k) = \alpha_{j_k} \xrightarrow{k} a$. Thus by (A′), (A) is satisfied. □

3.2 Notation: Let p be a fixed prime. Let Δ_p denote the space of all sequences $x = (x_0, x_1, \ldots)$ where $x_i \in \{0, 1, \ldots, p-1\}$. Give Δ_p the induced topology from the Cartesian product $\prod_{n=0}^{\infty} \{0, 1, \ldots, p-1\}_n$. For $x, y \in \Delta_p$, let n, m denote the least integers for which x_n and y_m are nonzero respectively. Let q be $\min(n, m)$. Define $x + y = z$ by the following:

(i) $z_i = 0$ for $i = 0, 1, \ldots, q - 1$

(ii) $x_q + y_q = pt_q + z_q$, $z_q \in \{0, 1, \ldots, p - 1\}$ where t_q is the uniquely defined integer

(iii) if $z_q, z_{q+1}, \ldots, z_k$ and $t_q, t_{q+1}, \ldots, t_k$ have been defined, then $x_{k+1} + y_{k+1} + t_k = pt_{k+1} + z_{k+1}$ (one is simply carrying).

Now Δ_p is a compact abelian group whose elements are called the p-adic integers, [HR, p. 109]. Its character group is $Z(p^\infty)$ [HR, p. 403], the discrete abelian group of all unimodular complex numbers of the form

$$t = \exp\left(2\pi i \frac{l}{p^{r+1}}\right), 0 \leq l < p^{r+1}.$$

The duality is given by

$$(x, t) = \exp\left(2\pi i \frac{l}{p^{r+1}} (x_0 + px_1 + \cdots + p^r x_r)\right).$$

The space Δ_p can be identified with the space of all infinite sequences $(k_n) = (k_1, k_2, \ldots)$ where $k_{n+1} \equiv k_n \pmod{p^n}$ and $0 \leq k_n < p^n$, [Ku, p. 154]. The correspondence $x \mapsto (k_n)$ is given by

$$k_{i+1} = x_0 + px_1 + \cdots + p^i x_i.$$

3.3 Theorem: *Let $G = \Delta_p$. Then $\mathscr{S} \not\subset \partial M$.*

Proof: We proceed as in Theorem 3.1. For $i = 1, 2, \ldots,$ let

$$V_i = p^i \Delta_p = \{(0, \ldots, 0, x_i, x_{i+1}, \ldots)\} \text{ and } x_{ij} = (0, \ldots, j, 0, \ldots)$$

where j appears in the $i - 1$ coordinate and $0 \leq j \leq p - 1$. Thus $m_i = p$ for all i.

Let $\pi \in \Delta$. We first calculate f_π^σ. Let $u = (1, 0, 0, \ldots)$. Then $\{0, \pm u, \pm 2u, \ldots\}$ is isomorphic to Z (expand $n \in Z_+$ to base p). Also for $n = jp^{i-1}$ $(0 \leq j \leq p - 1, i = 1, 2, \ldots)$, $x_{ij} = nu$. Since f_π^σ is a character on S, f_π^σ has the form $f_\pi^\sigma(nu) = \exp(2\pi i n\alpha)$ for some $\alpha \in [0, 1]$.

Let $\{b_n\}$ be a sequence of complex numbers with $|b_n|$ constant and nonzero. Suppose

$$\frac{2}{p(p+1)}\Big(p\,|b_n - b_{n-1}| + (p-1)\,|b_n \exp(2\pi i\alpha p^{n-1}) - b_{n-1}| + \cdots + |b_n \exp(2\pi i\alpha(p-1)p^{n-1}) - b_{n-1}|\Big) \xrightarrow{n} 0.$$

Thus $(b_n - b_{n-1}) \xrightarrow{n} 0$ and $(b_n \exp(2\pi i\alpha p^{n-1}) - b_{n-1}) \xrightarrow{n} 0$. Subtraction yields $b_n(1 - \exp(2\pi i\alpha p^{n-1})) \xrightarrow{n} 0$. Since $|b_n|$ is a nonzero constant, $\exp(2\pi i\alpha p^{n-1}) \xrightarrow{n} 1$. It follows that $\alpha = l/p^{r+1}$; that is, f_π^σ on S is the restriction of a continuous character. Thus by (B′), (B) holds.

Let
$$t = \exp\left(2\pi i \frac{l}{p^{r+1}}\right) \in Z(p^\infty)\ (0 < l < p^{r+1}).$$

Now $\hat{\mu}(t) = (\delta_1 * \cdots * \delta_{r+1})\hat{}(t)$ since $\operatorname{spt}(\mu_{r+1}) \subset V_{r+1}$ and $x \mapsto (x, t)$ is identically 1 on V_{r+1}. Thus

$$\hat{\mu}(t) = \prod_{n=1}^{r+1} \frac{2}{p(p+1)} \sum_{j=0}^{p-1} (p-j) \exp\left(2\pi i \frac{l}{p^{r+1}} j p^{n-1}\right).$$

We wish to show that $|\hat{\mu}(t)|$ is bounded away from 1. Each of the products above has absolute value ≤ 1, so we need show that only one of the terms are bounded in absolute value away from 1. Let $n = r + 1$, and look at

$$\alpha = \frac{2}{p(p+1)} \sum_{j=0}^{p-1} (p-j) \exp\left(2\pi i \frac{l}{p} j\right).$$

Let $w = \exp(2\pi i(l/p))$. Now w is a pth root of unity and the arguments given in the (C) part of Theorem 3.1 apply here to show that $|\alpha|$ is bounded away from 1. Thus (C) holds.

Now let
$$t_r = \exp\left(2\pi i \frac{1}{p^{r+1}}\right) \in Z(p^\infty).$$

Now
$$(x_{nj}, t_r) = \exp\left(2\pi i \frac{1}{p^{r+1}} j p^{n-1}\right) \xrightarrow{r} 1.$$

Thus $t_r \xrightarrow{r} 1$ on S. Also

$$\begin{aligned} \hat{\mu}(t_r) &= \prod_{n=1}^{r+1} \frac{2}{p(p+1)} \sum_{j=0}^{p-1} (p-j) \exp\left(2\pi i \frac{1}{p^{r+1}} j p^{n-1}\right) \\ &= \prod_{k=1}^{r+1} \frac{2}{p(p+1)} \sum_{j=0}^{p-1} (p-j) \exp\left(2\pi i j \frac{1}{p^k}\right) \\ &\xrightarrow{r} \prod_{k=1}^{\infty} \frac{2}{p(p+1)} \sum_{j=0}^{p-1} (p-j) \exp\left(2\pi i j \frac{1}{p^k}\right). \end{aligned}$$

To show that the infinite product converges to some z with $0 < |z| < 1$ we need only show that

$$\sum_{k=1}^{\infty} \left| 1 - \frac{2}{p(p+1)} \sum_{j=0}^{p-1} (p-j) \exp\left(2\pi i j \frac{1}{p^k}\right) \right| < \infty, \text{[RRC, p. 292]}.$$

This sum will be finite provided that

$$\sum_{k=1}^{\infty} \left| 1 - \exp\left(2\pi i \frac{j}{p^k}\right) \right| < \infty.$$

Now for large k,

$$\left|1 - \exp\left(2\pi i \frac{j}{p^k}\right)\right| = \left|2 \sin\left(\frac{\pi j}{p^k}\right)\right| \doteqdot \frac{2\pi j}{p^k}.$$

Thus since $\Sigma_{k=1}^{\infty} 1/p^k < \infty$, we have condition (A′) which yields (A). □

3.4 Theorem: *Let* $\mathbf{T}$ *be the additive group of reals modulo* 1. *Then* $\mathscr{S} \not\subset \partial M$.

Proof: Let $V_i = [0, 1/2^i)$, $i = 0, 1, \ldots$. For $i = 1, 2, \ldots$, let $x_{i0} = 0$ and $x_{i1} = 1/2^i$. Although the V_i's are not compact neighborhoods of 0, the previous calculations will apply if we show that μ_n is carried on V_n; that is, $\mu_n(\{1/2^n\}) = 0$. One first should note how to compute $\delta_1 * \cdots * \delta_n$:

(1) δ_1 lives on $\left\{0, \frac{1}{2}\right\}$ with weights $\left\{\frac{2}{3}, \frac{1}{3}\right\}$

(2) $\delta_1 * \delta_2$ lives on $\left\{0, \frac{1}{4}, \frac{1}{2}, \frac{3}{4}\right\}$ with weights $\left\{\frac{4}{3^2}, \frac{2}{3^2}, \frac{2}{3^2}, \frac{1}{3^2}\right\}$

(3) $\delta_1 * \delta_2 * \delta_3$ lives on $\left\{0, \frac{1}{2^3}, \frac{2}{2^3}, \frac{3}{2^3}, \frac{4}{2^3}, \frac{5}{2^3}, \frac{6}{2^3}, \frac{7}{2^3}\right\}$ with weights

$$\left\{\frac{8}{3^3}, \frac{4}{3^3}, \frac{4}{3^3}, \frac{2}{3^3}, \frac{4}{3^3}, \frac{2}{3^3}, \frac{2}{3^3}, \frac{1}{3^3}\right\}, \text{ etc.}$$

Thus we see that μ has mass $\frac{2}{3}$ on $[0, \frac{1}{2})$ and $\frac{1}{3}$ on $[\frac{1}{2}, 1)$; μ has mass $(\frac{2}{3})^2$ on $[0, \frac{1}{4})$; and in general, μ has mass $(\frac{2}{3})^n$ on $[0, 1/2^n)$. Hence

$$\mu\left(\left[\frac{m}{2^n}, \frac{m+1}{2^n}\right)\right) \leq \mu\left(\left[0, \frac{1}{2^n}\right)\right) \leq \left(\frac{2}{3}\right)^n \xrightarrow{n} 0.$$

It follows that μ and similarly μ_n have no discrete part; and so μ_n is concentrated (that is, carried) on V_n.

As before, we verify conditions (A′), (B′), and (C′). First note that for $\pi \in \Delta$, $f_\pi^\sigma(1/2^n)$ is a 2^nth root of unity. For (B′), let $\{b_n\}$ be a sequence of complex numbers with $|b_n|$ constant and nonzero. Assume that

$$\tfrac{2}{3}|b_n - b_{n-1}| + \tfrac{1}{3}|w_n b_n - b_{n-1}| \xrightarrow{n} 0$$

where w_n is a 2^nth root of unity. Thus $|b_n - b_{n-1}| \xrightarrow{n} 0$ which implies that $b_{n-1}/b_n \xrightarrow{n} 1$. Since

$$|b_n|\left|w_n - \frac{b_{n-1}}{b_n}\right| \xrightarrow{n} 0$$

and $|b_n|$ is constant, $w_n \xrightarrow{n} 1$. Now

$$w_n = f_\pi^\sigma\left(\frac{1}{2^n}\right) = f_\pi^\sigma\left(\frac{2}{2^{n+1}}\right) = \left(f_\pi^\sigma\left(\frac{1}{2^{n+1}}\right)\right)^2 = (w_{n+1})^2.$$

Since $w_n \xrightarrow{n} 1$, for n large, w_{n+1} is the principal square root of w_n. Thus there is an integer l such that $w_n = \exp(2\pi i(l/2^n))$ for n big, and thus for all n. It follows that on S, f_π^σ is the restriction of the character in $\hat{\mathbf{T}}$ given by $x \mapsto \exp(2\pi i l x)$. Hence (B) holds.

To show (C), we look at

$$\begin{aligned}\hat{\mu}(l) &= \int_0^1 \exp(2\pi i l x)\, d\mu(x) \\ &= \prod_{n=1}^{\infty} \left(\frac{2}{3} + \frac{1}{3}\exp\left(2\pi i \frac{l}{2^n}\right)\right).\end{aligned}$$

Now $|\hat{\mu}(l)| = 1$ if $l = 0$ and $\leq \frac{1}{3}$ otherwise (for l odd consider $n = 1$, and for $l = 2^k m$ with m odd consider $n = k + 1$). Thus the values of $|\hat{\mu}|$ are bounded away from 1 and (C) holds.

For (A), let $\gamma_k(x) = \exp(2\pi i 2^k x)$. Thus $\gamma_k(1/2^n) \xrightarrow{k} 1$. Hence $\gamma_k \xrightarrow{k} 1$ on S. Now

$$\begin{aligned}\hat{\mu}(\gamma_k) &= \prod_{n=1}^{\infty} \left(\frac{2}{3} + \frac{1}{3}\exp\left(2\pi i \frac{2^k}{2^n}\right)\right) \\ &= \prod_{r=1}^{\infty} \left(\frac{2}{3} + \frac{1}{3}\exp\left(2\pi i \frac{1}{2^r}\right)\right).\end{aligned}$$

This infinite product converges to $0 < |z| < 1$ since

$$\sum_{r=1}^{\infty} \left|1 - \frac{2}{3} - \frac{1}{3}\exp\left(2\pi i \frac{1}{2^r}\right)\right| = \frac{1}{3}\sum_{r=1}^{\infty} \left|1 - \exp\left(2\pi i \frac{1}{2^r}\right)\right| < \infty$$

as in Theorem 3.3. Thus (A) holds. □

3.5 Theorem: *Let* G = **R**, *the real line. Then* $\mathscr{S} \not\subset \partial M$.

Proof: One proceeds as in Theorem 3.4. For (B′), let $w_n = f_\pi^\sigma(1/2^n)$. As before, it follows that $w_n = \exp(2\pi i(l/2^n))$, $l \in \mathbf{R}$. The only difference in showing (C) is that here D_μ may be larger than before since $\hat{\mathbf{R}} \supset \hat{\mathbf{T}}$.

Let $a \in D_\mu$. Hence there is a net γ_α of continuous characters ($\gamma_\alpha(x) = \exp(2\pi i l_\alpha x)$) converging to $\mathbf{a}$ in $\sigma(L^\infty(\mu), L^1(\mu))$. Suppose $\{l_\alpha\}$ has a bounded subnet $\{l_\beta\}$ converging to $l \in \mathbf{R}$. Since the map of $l \mapsto \exp(2\pi i l x)$ from $\mathbf{R}$ to $(L^\infty(\mu), \sigma)$ is continuous (by the Lebesgue dominated convergence theorem), it follows that $\exp(2\pi i l_\beta x) \xrightarrow{\beta} \exp(2\pi i l x)$ in $\sigma(L^\infty(\mu), L^1(\mu))$. Thus $\mathbf{a} = \exp(2\pi i l x)$ in $L^\infty(\mu)$ and so $|a| = 1$. Otherwise, $\{l_\alpha\}$ has no bounded subnet. Thus we may assume $|l_\alpha| \geq 1$. For $2^k \leq |l_\alpha| \leq 2^{k+1}$, $k \geq 0$,

$$|\hat{\mu}(l_\alpha)| = \left|\int_{\mathbf{R}} \exp(2\pi i l_\alpha x)\, d\mu(x)\right| = \left|\prod_{n=1}^{\infty} \left(\frac{2}{3} + \frac{1}{3}\exp\left(\frac{2\pi i l_\alpha}{2^n}\right)\right)\right|$$

$$\leq \frac{1}{3}\left|2 + \exp\left(\frac{2\pi i l_\alpha}{2^{k+2}}\right)\right| \leq \frac{1}{3}\sqrt{5}$$

since the distance between the points 2 and $\exp(\pi i/2)$ is $\sqrt{5}$. Thus the values of $|\hat{\mu}|$ are bounded away from 1, and (C) holds. Part (A) is the same as in Theorem 3.4. □

4. Main Theorem

4.1 Lemma: *Let G be an* LCA *group and H a closed subgroup of G. Suppose there is a* $\mu \in M(H)$ *satisfying conditions* (A), (B), *and* (C) *for H. Then* μ, *considered as a measure on G, also satisfies* (A), (B), *and* (C). *Thus* $\mathscr{S} \not\subset \partial M$ *for G.*

Proof: Since μ is supported on H, (A) and (C) are clear. For (B), one need only to recall that the continuous characters on H can be extended to G, [R, p. 36]. □

4.2 Lemma: *Let G be an* LCA *group and H a compact subgroup. Suppose* $\mathscr{S}$ *is not contained in* ∂M *for* $M(G/H)$. *Then the same is true for* $M(G)$.

Proof: Let m_H be the Haar measure on H. Now m_H and $\delta - m_H$ (δ the unit in $M(G)$) are idempotents in $M(G)$. Let $A_1 = m_H * M(G)$ and $A_2 = (\delta - m_H) * M(G)$. Then $M(G) \cong A_1 \oplus A_2$ and $\Delta = \Delta(A_1) \cup \Delta(A_2)$, [Ri, p. 118]. The Shilov boundary of $M(G)$ is the union of the Shilov boundaries of A_1 and A_2. The same is true for the symmetric maximal ideals. This suffices since $A_1 \cong M(G/H)$ [R, p. 53; see 9.2.4 and 9.2.6]. □

4.3 Lemma: *Let G be an infinite abelian group. Then G contains as a subgroup,* Λ, *one of the following:*

(i) Z
(ii) $Z(p^\infty)$
(iii) *an infinite sum of cyclic groups.*

Proof: By way of contradiction, let us suppose not. Divisible abelian groups contain either a copy of the rationals, Q, or $Z(p^\infty)$, so G is a reduced group. Now G is a torsion group; that is, $G = \Sigma_p \oplus G_p$ where G_p is the subset of G consisting of the elements of order a power of the prime p. Since G is reduced, so is each G_p. Since each G_p has a finite cyclic direct summand [Ka, p. 21], G is the direct sum of a finite number of reduced primary groups, $G_{p_1}, \ldots, G_{p_n}$; one of which is infinite, call it Λ_1. Let Λ be a basic subgroup for Λ_1 [HR, p. 449]; that is, Λ is isomorphic with a sum of cyclic groups, Λ is a pure subgroup of Λ_1, and Λ_1/Λ is divisible. We claim that Λ is an infinite sum of cyclic groups. If not, then Λ is a pure subgroup of bounded order; and thus Λ is a direct summand of Λ_1 [Ka, p. 18]. Thus $\Lambda_1 \cong \Lambda \oplus (\Lambda_1/\Lambda)$. Since Λ_1 is reduced and Λ_1/Λ is divisible, we have that $\Lambda_1 = \Lambda$. It follows that Λ_1 is also finite which is the required contradiction. □

4.4 Main Theorem: *Let G be a nondiscrete* LCA *group. Then $\mathscr{S} \not\subset \partial M$.*

Proof: The principal structure theorem for LCA groups [HR, p. 389] yields that either G has a copy of $\mathbf{R}^n$, $n \geq 1$, in it or else G contains a compact open subgroup. In the first case we are done by Theorem 3.5 and Lemma 4.1. Thus by Lemma 4.1, it suffices to let G be a compact nondiscrete LCA group. We need only show that G contains a compact subgroup H such that $\mathscr{S} \not\subset \partial M$ for G/H. This will suffice by Lemma 4.2. Such an H will exist if $\hat{G}$ has a subgroup, Λ, which is the dual of a group with $\mathscr{S} \not\subset \partial M$ (since $\hat{\Lambda} = G/H$ where H is the annihilator of Λ). Thus we are done if $\hat{G}$ contains a direct sum of an infinite number of cyclic groups, if $\hat{G}$ contains a group $Z(p^\infty)$, or if $\hat{G}$ contains an element of infinite order (that is, $\hat{G}$ contains a copy of Z). This is the case by Lemma 4.3. □

4.5 Remark: To prove Theorem 4.4, one could try to show that on any infinite LCA group there is a $\mu \geq 0$ satisfying (A), (B), and (C). The result would then follow by Theorem 2.4. This is indeed possible and is given in Johnson [2]. One assumes G is an infinite compact abelian group with a subgroup, H, such that there is a $\mu \geq 0$ in $M(G/H)$ satisfying (A), (B), (C). Then one shows that the canonical measure on G defined by smearing μ over G works.

4.6 Remark: We now argue that $\Delta \neq \mathscr{S} \cup \partial M$. Let $\pi' \in \mathscr{S} \backslash \partial M$. Thus there is an open neighborhood, W of π' with $W \cap \partial M = \emptyset$. If $W \subset \mathscr{S}$, the Rossi local peak point theorem [Hö, p. 74] would imply that $\pi' \in \partial M$.

5. Historical Notes

5.1: J. Taylor [2] showed that $\partial M \neq \Delta$ for $G = \prod_{n=1}^{\infty} Z(2)_n$. The general result is due to Johnson whom we have closely followed [3]. Lemma 4.3 was pointed out to us by J. Oppelt.

Simon [1] showed that $\overline{\Gamma} \neq \mathscr{S}$ for $G = \mathbf{R}$, the real line. Johnson [2] extended this result to nondiscrete LCA groups. This result had been known to Varopoulos [1]. His proof, as was Simon's, is based on the existence of a positive measure in $M_0(G)$ (the algebra of measures with Fourier–Stieltjes transforms tending to 0 at infinity) whose support is compact and such that the subgroup generated by the support has Haar measure zero.

Lemma 2.5 is due to Williamson [1, p. 200]. That $\overline{\Gamma} \neq \partial M$ for G nondiscrete is shown in Williamson [1, p. 205]; also Rudin [3, p. 234].

CHAPTER 3

UNIFORM APPROXIMATION BY FOURIER-STIELTJES TRANSFORMS

1. Introduction

1.1: Let G be an infinite LCA group and $M(G)\hat{}$ the algebra of Fourier–Stieltjes transforms on the character group of G, $\Gamma(= \hat{G})$. We first show that $M(G)\hat{}$ is not closed in the topology of uniform convergence on Γ [Corollary 2.2]. We then seek to characterize its closure, $M(G)\hat{}^{-}$, that is, the algebra of all functions on Γ which can be approximated uniformly on Γ by Fourier–Stieltjes transforms. Our method is to pair $M(G)\hat{}$ and $M(\Gamma)$ in the natural way. One characterization follows easily from the Grothendieck completeness theorem [Theorem 3.6]. This characterization is, however, not easy to apply. We introduce the strict topology on $M(\Gamma)$ to get a characterization [Theorem 3.12, (A) and (B)] which is workable. This result will be needed in Chapter 4. Our method also yields a characterization of $M(G)\hat{}$ and the continuous almost periodic functions on Γ.

1.2 Remark: Let G be an LCA group and $\mu \in M(G)$. The Fourier–Stieltjes transform of μ, $\hat{\mu}$, is a complex-valued function on the character group of G, Γ, defined by $\hat{\mu}(\gamma) = \int_G \gamma(x)\, d\mu(x)$, $\gamma \in \Gamma$. Thus $\hat{\mu}$ is the restriction of the Gelfand transform to the set of uniqueness Γ. The collection of all $\hat{\mu}$ is denoted by $M(G)\hat{}$.

1.3 Remarks: (a) Each $\hat{\mu} \in M(G)\hat{}$ is a bounded, uniformly continuous function, in fact $\| \hat{\mu} \|_\infty \leq \| \mu \|$ (the uniform continuity follows from the regularity of μ).

(b) The map $\mu \mapsto \hat{\mu}$ is a complex homomorphism; that is, $(\mu * \nu)\hat{}(\gamma) = \hat{\mu}(\gamma)\hat{\nu}(\gamma)$.

(c) $M(G)\hat{}$ is invariant under translation; that is, $(\gamma\, d\mu)\hat{}(\gamma_0) = \hat{\mu}(\gamma_0 + \gamma)$ and for $\lambda(E) = \mu(E - x)$, $\hat{\lambda}(\gamma) = \gamma(x)\hat{\mu}(\gamma)$.

(d) $M(G)\hat{}$ is invariant under complex conjugation; that is, for

$$\mu^*(E) = \overline{\mu(-E)},\ \bar{\hat{\mu}} = \hat{\mu^*},\ [\text{R, p. 15}].$$

(e) If $\mu \in M(G)$ is such that $\hat{\mu} = 0$, then $\mu = 0$. This follows since μ is a regular measure and the trigonometric polynomials are uniformly dense in the continuous functions on compact sets [R, p. 17].

2. B*-subalgebras of M(G)

2.1 Theorem: *Let B be a commutative B^*-algebra such that B is algebraically*-isomorphic to a subalgebra of $M(G)$ (or $M(G)\hat{}$). Then B is finite dimensional.*

Proof: Let T be an algebraic $*$-isomorphism of B onto $B' \subset M(G)$. Since $M(G)$ is semisimple, the map is continuous [L, p. 76]. The map is topological since the sup-norm is the minimal norm in a B^*-algebra [Ri, p. 176].

Suppose B is infinite dimensional. Take any infinite linearly independent set, $\{f_n\}_{n=1}^{\infty}$, in B. Let B_0 be the self-adjoint closed subalgebra of B generated by $\{f_n\}_{n=1}^{\infty}$. Then B_0 is a separable B^*-algebra. Let B_0' be the image of B_0 in $M(G)$. Thus B_0' is a closed separable subspace of $M(G)$. Let $\{\mu_n\}_{n=1}^{\infty}$ be a set of probability measures with dense span in B_0'. Let $\mu = \sum_{n=1}^{\infty} (1/2^n)\mu_n$. It follows that $B_0' \subset L^1(\mu)$, and so B_0' is sequentially weakly complete [DS, p. 290]. Thus B_0 is sequentially weakly complete. But $B_0 \cong C_0(S)$, the space of complex-valued continuous functions vanishing at infinity on the maximal ideal space of B_0. But $C_0(S)$ is not sequentially weakly complete unless finite dimensional [DS, p. 340]. The last assertion proceeds as follows. Since B_0 is separable, S is metrizable. The weak sequential completeness of $C_0(S)$ implies that the characteristic functions of points are in $C_0(S)$. Thus S is discrete. Since $C_0(S)$ is separable, S is σ-compact. The weak sequential completeness of $C_0(S)$ implies that the constant function 1 is in $C_0(S)$. Thus S is discrete and finite and $C_0(S)$ is therefore finite dimensional. □

2.2 Corollary: *Let $A \subset \Gamma$ be an open set and $C_0(A)$ the algebra of continuous functions on Γ which vanish at infinity and which vanish off A. Then $C_0(A)$ is contained in $M(G)\hat{}$ if and only if A is finite. In particular, $L^1(G)\hat{}$ is a proper subset of $C_0(\Gamma)$ (Γ is infinite), and $M(G)\hat{}$ is not closed in the uniform topology on Γ.*

2.3 Corollary: *Let G be an infinite* LCA *group and $AP(\Gamma)$ the algebra of continuous almost periodic functions on Γ. Then $AP(\Gamma)$ is not contained in $M(G)\hat{}$.*

Proof: Let G be an infinite LCA group. Then Γ is also infinite and $AP(\Gamma)$ is an infinite dimensional commutative B^* algebra [L, p. 166, 170]. □

2.4 Corollary: *Let Γ be a nondiscrete* LCA *group and U an open set in Γ. Then there exists a continuous function f vanishing off U such that $f \notin M(G)\hat{}$.*

Proof: If not, then Corollary 2.2 implies that U is a finite set and thus Γ would be discrete. □

2.5 Corollary: *If every continuous function on Γ belongs locally to $M(G)\hat{}$, then Γ is discrete.*

Proof: We first note that the converse is trivial.

Let $C_c(\Gamma)$ denote the continuous complex-valued functions on Γ with compact support. Let $f \in C_c(\Gamma)$ and spt $(f) = K$. We will show that $C_c(\Gamma) \subset L^1(G)\hat{} \subset M(G)\hat{}$. For each $x \in K$, let $\hat{\mu_x} \in M(G)\hat{}$ be such that $\hat{\mu_x} = f$ on an open neighborhood, V_x, of x. Since K is compact, we can find $x_1, \ldots, x_n$ such that $V_{x_1}, \ldots, V_{x_n}$ is a cover of K. Let $\hat{h_1}, \ldots, \hat{h_n}$ be chosen in $L^1(G)\hat{}$ such that $\hat{h_i}$ vanishes outside of V_{x_i}, $1 \leq i \leq n$, and such that $\sum_{i=1}^n \hat{h}_i(x) = 1$ on K [R, p. 49 and RRC, p. 40]. Thus for all $x \in \Gamma$,

$$f(x) = f(x) \sum_{i=1}^{n} \hat{h}_i(x) = \sum_{i=1}^{n} f(x)\hat{h}_i(x) = \sum_{i=1}^{n} \hat{\mu}_{x_i}(x)\hat{h}_i(x)$$

$$= \sum_{i=1}^{n} (\mu_{x_i} * h)\hat{}(x) = \hat{v}(x)$$

where $v = \sum_{i=1}^n \mu_{x_i} * h$. Thus $C_c(\Gamma) \subset L^1(G)\hat{}$. The result follows now from Corollary 2.4. □

3. Pairings with M(G)

3.1 Definition: For $\hat{\mu} \in M(G)\hat{}$ and $\lambda \in M(\Gamma)$, define $\langle \hat{\mu}, \lambda \rangle = \int_\Gamma \hat{\mu}\, d\lambda$ ($= \int_G \hat{\lambda}\, d\mu$ by Fubini's Theorem). Let w denote the weak topology on $M(\Gamma)$ from this pairing; that is, in w, $\lambda_\alpha \xrightarrow{\alpha} 0$ if and only if $\langle \hat{\mu}, \lambda_\alpha \rangle \xrightarrow{\alpha} 0$ for all $\hat{\mu} \in M(G)\hat{}$. Let B_n denote the n-ball of $M(\Gamma)$ and $C_n = \{\lambda \in B_n : \hat{\lambda} \in C_0(G)\}$. Let $\mathscr{T}_{B_n}$ denote the topology on $M(G)\hat{}$ of uniform convergence on the norm balls, B_n; that is, $\hat{\mu_\alpha} \xrightarrow{\alpha} 0$ in $\mathscr{T}_{B_n}$ if and only if $\sup \{ |\langle \hat{\mu_\alpha}, \lambda \rangle| : \lambda \in B_n\} \xrightarrow{\alpha} 0$ for all n.

3.2 Lemma: *$\langle \cdot, \cdot \rangle$ is a pairing.*

Proof: If $\langle \hat{\mu}, \lambda \rangle = 0$ for all $\hat{\mu} \in M(G)\hat{}$, in particular for $\hat{\mu} \in C_0(\Gamma)$, then $\lambda = 0$ since $M(G)\hat{} \cap C_0(\Gamma)$ is sup-norm dense in $C_0(\Gamma)$, and similarly if $\langle \hat{\mu}, \lambda \rangle = 0$ for all $\lambda \in M(G)$. □

3.3 Lemma: *B_n is w-closed.*

Proof: Let $\{\lambda_\alpha\} \subset B_n$ be such that $\lambda_\alpha \xrightarrow{\alpha} \lambda$ in w, $\lambda \in M(\Gamma)$. We wish to show that $\| \lambda \| \leq n$. Now $\langle \hat{\mu}, \lambda_\alpha \rangle \xrightarrow{\alpha} \langle \hat{\mu}, \lambda \rangle$ for $\hat{\mu} \in M(G)\hat{} \cap C_0(\Gamma)$. If $\| \hat{\mu} \|_\infty \leq 1$, then

$$|\langle \hat{\mu}, \lambda_\alpha \rangle| \leq \| \hat{\mu} \|_\infty \, \| \lambda_\alpha \| \leq n,$$

so $| \langle \hat{\mu}, \lambda \rangle | \leq n$. Since $M(G)\hat{} \cap C_0(\Gamma)$ is sup-norm dense in $C_0(\Gamma)$, $\| \lambda \| \leq n$. □

3.4 Lemma: *B_n is w-bounded.*

Proof: For $\hat{\mu} \in M(G)\hat{}$, let $V = \{\lambda \in M(\Gamma) : | \langle \hat{\mu}, \lambda \rangle | \leq 1\}$. Now V is a typical subbasic w-neighborhood of 0. We wish to find $\rho > 0$ such that $B_n \subset \rho V$. Since $| \langle \hat{\mu}, \lambda \rangle | \leq \| \hat{\mu} \|_\infty \, \| \lambda \|$, we let $\rho = n \, \| \hat{\mu} \|_\infty$. □

3.5 Lemma: *On $M(G)\hat{}$, the topology, $\mathscr{T}_{B_n}$, of uniform convergence on the norm balls, B_n, is equivalent to the sup-norm topology on Γ.*

Proof: Let $\{\hat{\mu}_\alpha\} \subset M(G)\hat{}$. Suppose $\hat{\mu}_\alpha \xrightarrow{\alpha} 0$ in $\mathscr{T}_{B_n}$, then $\sup \{ | \langle \hat{\mu}_\alpha, \lambda \rangle | : \lambda \in B_n\} \xrightarrow{\alpha} 0$. Hence $\| \hat{\mu}_\alpha \|_\infty = \sup \{ | \langle \hat{\mu}_\alpha, \lambda \rangle | : \lambda \in L^1(\Gamma)$ with $\| \lambda \|_1 \leq 1\} \xrightarrow{\alpha} 0$.

Suppose $\| \hat{\mu}_\alpha \|_\infty \xrightarrow{\alpha} 0$. Then $\sup \{ | \langle \hat{\mu}_\alpha, \lambda \rangle | : \lambda \in B_n\} \leq \| \hat{\mu}_\alpha \|_\infty \cdot n \xrightarrow{\alpha} 0$. □

3.6 Theorem: *For $f \in C^B(\Gamma)$, the space of continuous bounded functions on Γ, the following are equivalent:*

(A) $f \in M(G)\hat{}^{-}$.

(C) *The linear functional on $M(\Gamma)$ determined by f, that is, $\lambda \mapsto \int_\Gamma f \, d\lambda$, is w-continuous when restricted to the norm balls, B_n, of $M(\Gamma)$.*

Proof: The norm balls, B_n, are convex, circled, w-closed, w-bounded, and $\bigcup_{n=1}^\infty B_n = M(\Gamma)$. We now are able to apply the Grothendieck completion theorem [Ko, p. 271; KN, p. 145] to characterize the $\mathscr{T}_{B_n}$ (or sup-norm) completion of $M(G)\hat{}$, $M(G)\hat{}^{-}$, as the linear functionals on $M(\Gamma)$ which are w-continuous on the norm balls, B_n, of $M(\Gamma)$.

Let $M(\Gamma)'$ denote the space of all linear functionals on $M(\Gamma)$. It remains to show that the correspondence between $f \in M(G)\hat{}^{-}$ and $F \in M(\Gamma)'$ with

F w-continuous on B_n is given by $F(\lambda) = \int_\Gamma f\, d\lambda$. Now for $g \in C^B(\Gamma)$ we associate a linear functional on $M(\Gamma)$ by $\lambda \mapsto \int_\Gamma g\, d\lambda$. Thus, via this association, we may view $C^B(\Gamma)$ as a subspace of $M(\Gamma)'$. The arguments of Lemma 3.5 show that on $C^B(\Gamma)$ ($\subset M(\Gamma)'$), $\mathscr{T}_{B_n}$ is equivalent to the sup-norm topology on Γ. Thus $f \in M(G)\hat{}^{-} \subset C^B(\Gamma)$ corresponds to $F \in M(\Gamma)'$ with F w-continuous on B_n where $F(\lambda) = \int_\Gamma f\, d\lambda$. □

3.7 Definition: Represent the functions, f, in $C^B(G)$ as operators, T_f, on $C_0(G)$ by $T_f(g) = fg$, $g \in C_0(G)$. Let WO be the weak operator topology on $C^B(G)$ via this representation and SO be the strong operator topology. Let $\{f_\alpha\} \subset C^B(G)$:

(i) $f_\alpha \xrightarrow{\alpha} 0$ in WO if and only if for $g \in C_0(G)$ and $\mu \in M(G)$, $\int_G f_\alpha g\, d\mu \xrightarrow{\alpha} 0$.

(ii) $f_\alpha \xrightarrow{\alpha} 0$ in SO if and only if for $g \in C_0(G)$, $\| f_\alpha g \|_\infty \xrightarrow{\alpha} 0$.

Viewing $M(\Gamma)$ as a subalgebra of $C^B(G)$ via the Fourier-Stieltjes transformation induces the WO and SO topologies on $M(\Gamma)$.

3.8 Lemma: *Let $S \subset C^B(G)$ be a sup-norm bounded set. Then on S the SO topology is equivalent to the compact-open topology, κ.*

Proof: Let $S \subset C^B(G)$ be such that $\| f \|_\infty \leq M$ for all $f \in S$. Let f_0 be in the κ-closure of S. Given $\varepsilon > 0$ and any $g \in C_0(G)$, choose a compact set $K \subset G$ such that $|g(x)| < \varepsilon$ for $x \in G \backslash K$. Then for $f \in S$ with $\| f - f_0 \|_K < \varepsilon$ (where $\| f \|_K = \sup\{|f(x)| : x \in K\}$), we have

$$\begin{aligned} \| g(f - f_0) \|_\infty &\leq \| g(f - f_0) \|_K + \| g(f - f_0) \|_{G\backslash K} \\ &\leq \| f - f_0 \|_K \| g \|_\infty + \varepsilon \| f - f_0 \|_{G\backslash K} \\ &\leq \varepsilon \| g \|_\infty + \varepsilon(M + \| f_0 \|_\infty) \\ &= \varepsilon(\| g \|_\infty + M + \| f_0 \|_\infty). \end{aligned}$$

Thus f_0 is in the SO-closure of S. Thus $SO \subset \kappa$.

In general $\kappa \subset SO$. This follows since for $\{f_\alpha\} \subset C^B(G)$, $f_\alpha \xrightarrow{\alpha} 0$ in κ if and only if $\| f_\alpha g \|_\infty \xrightarrow{\alpha} 0$ for $g \in C_c(G)$, the continous functions on G with compact support. □

3.9 Lemma: *On the norm balls, B_n, the weak topology, w, is equivalent to the weak operator topology, WO. In general, $WO \subset w$.*

Proof: Let $\{\lambda_\alpha\} \subset M(\Gamma)$. Suppose $\lambda_\alpha \xrightarrow{\alpha} 0$ in w, that is, $\langle \hat{\mu}, \lambda_\alpha \rangle = \int_\Gamma \hat{\mu}\, d\lambda_\alpha = \int_G \hat{\lambda}_\alpha\, d\mu \xrightarrow{\alpha} 0$, for each $\mu \in M(G)$. For $g \in C_0(G)$ and $\mu \in M(G)$, $g\, d\mu \in M(G)$. So $\int_G \hat{\lambda}_\alpha g\, d\mu \xrightarrow{\alpha} 0$. Hence $\lambda_\alpha \xrightarrow{\alpha} 0$ in WO.

Suppose $\lambda_\alpha \xrightarrow{\alpha} 0$ in WO in B_n, that is, $\| \lambda_\alpha \| \leq n$ and $\int_G \hat{\lambda}_\alpha g\, d\mu \xrightarrow{\alpha} 0$ for $g \in C_0(G)$ and $\mu \in M(G)$. Let $\mu \in M(G)$ and $\varepsilon > 0$. There exists a compact subset, K, of G such that $\| \mu | K - \mu \| \leq \varepsilon/3n$. Let $g \in C_0(G)$ be such that

$g = 1$ on K and $\| g \|_\infty \leq 1$. Let α_0 be such that if $\alpha \geq \alpha_0$, then $| \int_G \hat{\lambda_\alpha} g \, d\mu | \leq \varepsilon/3$. Thus for $\alpha \geq \alpha_0$,

$$\begin{aligned} | \textstyle\int_G \hat{\lambda_\alpha} \, d\mu | &\leq | \textstyle\int_K \hat{\lambda_\alpha} \, d\mu | + | \textstyle\int_{G\backslash K} \hat{\lambda_\alpha} \, d\mu | \\ &\leq | \textstyle\int_K \hat{\lambda_\alpha} g \, d\mu | + \| \hat{\lambda_\alpha} \|_\infty \, \| \mu | G\backslash K \| \\ &\leq | \textstyle\int_G \hat{\lambda_\alpha} g \, d\mu | + | \textstyle\int_{G\backslash K} \hat{\lambda_\alpha} g \, d\mu | + n \frac{\varepsilon}{3n} \\ &\leq \frac{\varepsilon}{3} + \| \hat{\lambda_\alpha} g \|_\infty \, \| \mu | G\backslash K \| + \frac{\varepsilon}{3} \\ &\leq \frac{\varepsilon}{3} + n \frac{\varepsilon}{3n} + \frac{\varepsilon}{3} = \varepsilon. \quad \square \end{aligned}$$

3.10 Lemma: *Let* $\{\lambda_n\}_{n=1}^\infty \subset B_m$. *If* $\hat{\lambda_n}(x) \xrightarrow{n} 0$ *for all* $x \in G$, *then* $\lambda_n \xrightarrow{n} 0$ *in* w, *and vice versa.*

Proof: Let $\{\lambda_n\} \subset M(\Gamma)$ be such that $\| \lambda_n \| \leq m$ and $\hat{\lambda_n}(x) \xrightarrow{n} 0$ for all $x \in G$. We wish to show that $\lambda_n \xrightarrow{n} 0$ in w; that is, let $\hat{\mu} \in M(G)\hat{}$ and show $\langle \hat{\mu}, \lambda_\alpha \rangle \xrightarrow{\alpha} 0$. Let $\varepsilon > 0$ and K be a compact subset of G such that $\| \mu - \mu | K \| \leq \varepsilon/2m$. Since $\| \hat{\lambda_n} \|_\infty \leq m$, $\{\hat{\lambda_n} | K\}$ are sup-norm bounded and pointwise convergent to 0. Therefore, $\int_K \hat{\lambda_n} \, d\mu \xrightarrow{n} 0$ by the Lebesgue dominated convergence theorem; hence there exists n_0 such that if $n \geq n_0$, then $| \int_K \hat{\lambda_n} \, d\mu | \leq \varepsilon/2$. Hence for $n \geq n_0$,

$$\begin{aligned} | \langle \hat{\mu}, \lambda_n \rangle | &= | \textstyle\int_\Gamma \hat{\mu} \, d\lambda_n | = | \textstyle\int_G \hat{\lambda_n} \, d\mu | \\ &\leq | \textstyle\int_K \hat{\lambda_n} \, d\mu | + | \textstyle\int_{G\backslash K} \hat{\lambda_n} \, d\mu | \leq \frac{\varepsilon}{2} + \| \hat{\lambda_n} \|_\infty \, \| \mu | G\backslash K \| \\ &\leq \frac{\varepsilon}{2} + m \frac{\varepsilon}{2m} = \varepsilon. \end{aligned}$$

The converse is clear by considering unit point measures. $\square$

3.11 Lemma: *A complex linear functional on* $M(\Gamma)$ *is SO-continuous on the circled convex set C if and only if it is WO-continuous on C.*

Proof: We are regarding $M(\Gamma)$ as a subset of the bounded linear operators on $C_0(G)$, $\mathscr{B}(C_0(G))$, by $T_\mu(f) = \hat{\mu} f$. The crucial property we now use is that convex subsets of $\mathscr{B}(C_0(G))$ have the same *SO* and *WO* closures in $\mathscr{B}(C_0(G))$ [DS, p. 477].

Since $WO \subset SO$, a *WO*-continuous linear functional on C is *SO*-continuous on C.

Let Φ be a linear functional on $M(\Gamma)$ which is *SO*-continuous on C.

Extend Φ to $\mathscr{B}(C_0(G))$ and let K denote its kernel. Now Φ being SO-continuous on C implies that $K \cap C$ is SO-closed in C [KN, p. 113] and hence $\overline{K \cap C}^{SO} \cap C = K \cap C$ and hence $\overline{K \cap C}^{WO} \cap C = K \cap C$ [DS, p. 477] and hence $K \cap C$ is WO-closed in C and hence Φ is WO-continuous on C. □

3.12 Theorem: *Let $f \in C^B(\Gamma)$. The following are equivalent:*

(A) $f \in M(G)\hat{}^{-}$.

(B) *If* $\{\lambda_n\}_{n=1}^{\infty} \subset M(\Gamma)$, $\| \lambda_n \| \leq m$, *and* $\hat{\lambda}_n(x) \xrightarrow{n} 0$ *for all* $x \in G$, *then* $\int_\Gamma f\, d\lambda_n \xrightarrow{n} 0$.

(C) $\lambda \mapsto \int_\Gamma f\, d\lambda$ *is w-continuous on* B_n.

(D) $\lambda \mapsto \int_\Gamma f\, d\lambda$ *is SO-continuous on* B_n.

(E) $\lambda \mapsto \int_\Gamma f\, d\lambda$ *is WO-continuous on* B_n.

Proof: (A) is equivalent to (C) by Theorem 3.6.
(C) is equivalent to (E) by Lemma 3.9.
(E) is equivalent to (D) by Lemma 3.11.
(C) implies (B) by Lemma 3.10.

It remains to show that (B) implies (A). We first note that if G is σ-compact, then the compact-open topology is metrizable and it is immediate that (B) implies (D) which implies (A).

We show now that $\lambda \mapsto \int_\Gamma f\, d\lambda$ is SO-continuous on $C_m = \{\lambda \in B_m: \hat{\lambda} \in C_0(G)\}$. For suppose not. Then there would exist $\varepsilon > 0$ such that for any compact $K \subset G$ and $\delta > 0$ we have $\lambda_{K,\delta}$ in $V_{K,\delta} = \{\lambda \in C_m : \| \hat{\lambda} \|_K \leq \delta\}$ such that $| \int_\Gamma f d\lambda_{K,\delta} | \geq \varepsilon$. Let λ_1 be any measure in C_m such that $| \int_\Gamma f d\lambda_1 | \geq \varepsilon$.

For $K_1 = \{x \in G: | \hat{\lambda}_1(x) | \geq 1\}$, let $\lambda_2 = \lambda_{K_1, 1}$.

For $K_2 = \{x \in G: | \hat{\lambda}_i(x) | \geq \frac{1}{2}$, some $i = 1, 2\}$, let $\lambda_3 = \lambda_{K_2, \frac{1}{2}}$.

For $K_n = \{x \in G: | \hat{\lambda}_i(x) | \geq 1/n$, some $i = 1, 2, \ldots, n\}$, let $\lambda_{n+1} = \lambda_{K_n, 1/n}$.

Now $\hat{\lambda}_n(x) \xrightarrow{n} 0$ for all $x \in G$ and $\{\lambda_n\} \subset C_m \subset B_m$. Hence (B) implies that $\int_\Gamma f\, d\lambda_n \xrightarrow{n} 0$. But $| \int_\Gamma f\, d\lambda_n | \geq \varepsilon$. This contradiction shows that $\lambda \mapsto \int_\Gamma f\, d\lambda$ is SO-continuous on the norm balls of $M_0(\Gamma) = \{\mu \in M(\Gamma): \hat{\mu} \in C_0(G)\}$.

Let $(B_0) - (E_0)$ denote statements (B) – (E) with $M(\Gamma)$ and B_n replaced by $M_0(\Gamma)$ and C_n. We have now shown that (B) implies (D_0). Because of the application of this theorem given in the next chapter, we have paired $M(G)\hat{}$ with $M(\Gamma)$. However, one could have paired $M(G)\hat{}$ with $M_0(\Gamma)$. In 3.2–3.6, one can replace $M(\Gamma)$ and B_n with $M_0(\Gamma)$ and C_n. Thus our arguments above show that (A) is equivalent to $(B_0) - (E_0)$; in particular, (B) which implies (D_0) does imply (A). □

3.13 Corollary: *Let Z denote the set of integers and f a (continuous) bounded function on Z be such that $f(10^k + n) = e^{-in^{\frac{1}{2}}}$ $(k = 1, 2, \ldots, 1 \leq n \leq k)$. Then $f \notin M(\mathbf{T})\hat{}^{-}$ where $\mathbf{T}$ is the unit circle, the character group of Z.*

Proof: Let

$$\lambda_k = \frac{1}{k} \sum_{n=1}^{k} e^{in^{\frac{1}{2}}} \delta_{10k+n},$$

where δ_p is unit point mass at p. Then $\| \lambda_k \| = 1$. It is known [Z, p. 200] that for each $x \in \mathbf{T}$, there is $C_x > 0$ such that

$$\left| \sum_{n=1}^{k} e^{in^{\frac{1}{2}}} e^{inx} \right| \leq C_x \cdot k^{3/4}.$$

Hence

$$|\hat{\lambda_k}(x)| = \left| \frac{1}{k} \sum_{n=1}^{k} e^{in^{\frac{1}{2}}} e^{i(10^k+n)x} \right| = \left| \frac{1}{k} \sum_{n=1}^{k} e^{in^{\frac{1}{2}}} e^{inx} \right| \leq \frac{1}{k} C_x k^{3/4} \xrightarrow{k} 0.$$

By Theorem 3.12, $f \in M(\mathbf{T})\hat{}^{-}$ implies that $\Sigma_Z f\, d\lambda_k \xrightarrow{k} 0$. But

$$\sum_Z f\, d\lambda_k = \frac{1}{k} \sum_{n=1}^{k} 1 = 1 \overset{k}{\not\to} 0.$$

Hence $f \notin M(\mathbf{T})\hat{}^{-}$. □

3.14 Remark: Theorem 3.12 remains valid if $M(\Gamma)$ is replaced by $L^1(\Gamma)$. The proof is essentially the same except that the final argument in Theorem 3.12 is unnecessary.

We now show how the technique used to characterize $M(G)\hat{}^{-}$ yields a characterization of the continuous almost periodic functions on Γ, $AP(\Gamma)$ [see L, p. 165]. The difference is a replacement of sequences in Theorem 3.12 (B) by nets.

3.15 Theorem: *Let G be an* LCA *group. For $f \in C^B(\Gamma)$ the following are equivalent:*

(a) $f \in AP(\Gamma)$.

(b) *If $\{\lambda_\alpha\} \subset B_n$ and $\hat{\lambda_\alpha} \xrightarrow{\alpha} 0$ pointwise on G, then $\int_\Gamma f\, d\lambda_\alpha \xrightarrow{\alpha} 0$.*

Proof: We pair $M(\Gamma)$ this time with $M_d(G)\hat{}$, the Fourier-Stieltjes transforms of discrete measures on G by $\langle \hat{\mu}, \lambda \rangle = \int_\Gamma \hat{\mu}\, d\lambda$. Let τ denote the weak topology on $M(\Gamma)$ from this pairing.

Now B_n is a convex circled set in $M(\Gamma)$. It is τ-bounded since

$$|\langle \hat{\mu}, \lambda \rangle| \leq \| \hat{\mu} \|_\infty \| \lambda \|.$$

It is τ-closed since

$$\sup\{ | \textstyle\int_\Gamma \hat{\mu}\, d\lambda | : \mu \in M_d(G), \| \hat{\mu} \|_\infty \leq 1\} = \| \lambda \|.$$

Let $\mathscr{T}_{B_n}$ denote the topology on $M_d(G)\hat{}$ of uniform convergence on the sets B_n. Since

$$\sup\{ | \textstyle\int_\Gamma \hat{\mu}\, d\lambda | : \lambda \in M(\Gamma) : \| \lambda \| \leq 1\} = \| \hat{\mu} \|_\infty,$$

$\mathscr{T}_{B_n}$ is the sup-norm topology.

The Grothendieck completeness theorem implies that for $f \in C^B(\Gamma)$, $f \in M_d(G)\hat{}^{-}$ if and only if $\lambda \mapsto \int_\Gamma f\, d\lambda_\alpha$ is τ-continuous on B_n; that is, $f \in M_d(G)\hat{}^{-}$ if and only if (b). But $M_d(G)\hat{}^{-} = AP(\Gamma)$ by the following lemma. $\square$

3.16 Lemma: $M_d(G)\hat{}^{-} = AP(\Gamma)$.

Proof: A function f is in $AP(\Gamma)$ if and only if f has a continuous extension to the almost periodic compactification, Γ^a; that is, $AP(\Gamma) \cong C(\Gamma^a)$ [L, p. 168]. Now Γ^a is the Bohr group of Γ [L, p. 173]. Now $M_d(G)\hat{}$ contains the continuous characters on Γ and they separate the points of Γ^a [R, p. 31]. Thus by the Stone-Weierstrass Theorem, $M_d(G)\hat{}^{-} \cong C(\Gamma^a) \cong AP(\Gamma)$. $\square$

3.17 Remark: Our method yields also a characterization of $M(G)\hat{}$. Let $M(G)\hat{}$ and $M(\Gamma)$ be paired as before. Let $\hat{B_n} = \{\lambda \in M(\Gamma): \| \hat{\lambda} \|_\infty \le n\}$ and $\hat{\mathcal{T}}$ the topology on $M(G)\hat{}$ of uniform convergence on the sup-norm balls $\hat{B_n}$ with respect to this pairing. $\hat{\mathcal{T}}$ is equivalent to the measure-norm topology on $M(G)\hat{}$ since, for $\hat{\mu} \in M(G)\hat{}$,

$$\| \mu \| = \frac{1}{n} \sup \{ | \langle \hat{\mu}, \lambda \rangle | : \lambda \in \hat{B_n} \}$$

and

$$| \langle \hat{\mu}, \lambda \rangle | \le \| \mu \| \, \| \hat{\lambda} \|_\infty .$$

The method for proving the following theorem is similar to that for Theorem 3.12.

3.18 Theorem: *Let $f \in C^B(\Gamma)$. The following are equivalent:*

(A′) $f \in M(G)\hat{}$.

(B′) *If* $\{\lambda_n\} \subset M(\Gamma)$, $\| \hat{\lambda_n} \|_\infty \le m$, *and* $\hat{\lambda_n}(x) \xrightarrow{n} 0$ *for all* $x \in G$, *then* $\int_\Gamma f\, d\lambda_n \xrightarrow{n} 0$.

(C′) $\lambda \mapsto \int_\Gamma f\, d\lambda$ *is w-continuous on* $\hat{B_n}$.

(D′) $\lambda \mapsto \int_\Gamma f\, d\lambda$ *is SO-continuous on* $\hat{B_n}$.

(E′) $\lambda \mapsto \int_\Gamma f\, d\lambda$ *is WO-continuous on* $\hat{B_n}$.

3.19 Remark: Let G be a nonabelian locally compact group, $A(G)$ the Fourier algebra (for G compact see 8.4.12, and G noncompact see Eymard [1]), and $VN(G)$ the von Neumann algebra of operators on $L^2(G)$ generated by the left translation operators (for G compact, $VN(G) \cong \mathcal{L}^\infty(\hat{G})$, see 8.3.2). Now $M(G)$ can be viewed in $VN(G)$ as left convolution operators on $L^2(G)$. For G abelian, $VN(G) \cong L^\infty(\hat{G})$ and Remark 3.14 characterizes the closure of $M(G)$ in $VN(G)$. Using the duality developed in Eymard [1, p. 210], each $T \in VN(G)$ corresponds to $\phi_T \in A(G)^*$ (see 8.4.17 for G compact). The nonabelian analogue of Remark 3.14 also holds: for $T \in VN(G)$, $T \in M(G)^-$ ($\subset VN(G)$) if and only if $f_n \in A(G)$, $\| f_n \|_{A(G)} \le 1$, and $f_n \xrightarrow{n} 0$ pointwise on G implies $\phi_T(f_n) \xrightarrow{n} 0$ (equivalently, $T(f_n) \xrightarrow{n} 0$ pointwise on G).

4. Historical Notes

4.1: That $M(G)$, G an infinite LCA group, does not contain an infinite dimensional commutative B^*-subalgebra (Theorem 2.1) was observed by Edwards [1, p. 72]. His result implies a theorem of Segal [1] (Corollary 2.2), and of Hewitt [1] (Corollary 2.3). Corollaries 2.4 and 2.5 are also due to Edwards [1].

The characterization of $M(G)\hat{}^{-}$ is due to Ramirez [1, p. 327]. The result for G and Γ σ-compact was reported earlier by Beurling and Hewitt [Hewitt 2, p. 138]. We have modified Lemma 3.11 following a suggestion of R. Burckel.

Our characterization of the continuous almost periodic functions was first shown by Edwards [2, p. 254]. We have followed the proof in Ramirez [5]. Edwards [2] has also obtained Theorem 3.6.

The SO-topology on $C^B(G)$ is called the strict topology and was introduced by Buck [1].

Theorem 3.18 appears in Ramirez [1].

CHAPTER 4

WEAKLY ALMOST PERIODIC FUNCTIONS

1. Introduction

1.1: In this chapter, we show that $M(\Gamma)\hat{}^{-} \subset WAP(G)$ with equality only when G is compact.

1.2 Definition: Let G be a locally compact abelian group. Let $f \in C^B(G)$. The orbit of f, $O(f)$, denotes the set of translates of f, $\{f_s : s \in G\}$ where $f_s(t) = f(s + t)$. We call f **weakly almost periodic** (w.a.p.) if the orbit of f is relatively weakly compact. We denote by $WAP(G)$ the space of weakly almost periodic functions on G.

1.3 Remark: One of the important properties of weakly almost periodic functions (first observed by Grothendieck [1, p. 183]) is that for any two sequences $\{t_n\}$, $\{s_m\} \subset G$ the $\lim_m \lim_n f(t_n + s_m) = \lim_n \lim_m f(t_n + s_m)$ whenever each of the limits exists. In fact, this characterizes the w.a.p. functions. To show that $f \in C^B(G)$ is w.a.p. one need only show that $O(f)$ is weakly sequentially compact or weakly countably compact.

For $f \in C^B(G)$, the following are equivalent:

(i) $O(f)$ is relatively weakly compact.

(ii) $O(f)$ is weakly countably compact, that is, each sequence in $O(f)$ has a weak cluster point in $C^B(G)$.

(iii) $O(f)$ is weakly sequentially compact, that is, each sequence in $O(f)$ has a subsequence which converges weakly in $C^B(G)$.

(iv) $\lim_m \lim_n f(t_n + s_m) = \lim_n \lim_m f(t_n + s_m)$ whenever each of the limits exists.

The equivalence of (i), (ii), and (iii) is contained in the Eberlein–Šmulian Theorem [DS, p. 430]. The equivalence of (iii) and (iv) follows from [KN, pp. 76–79; DS, p. 269]. (We can identify $C^B(G)$ with the continuous functions on the Stone–Čech compactification of G, βG. In $C^B(\beta G)$, $f_n \to f$ weakly if and only if the sequence is uniformly bounded and pointwise convergent [DS, p. 265].)

2. Basic Facts

2.1 Definition: Let $f \in C^B(G)$ be such that $\sum_{i,j} \alpha_i \bar{\alpha}_j f(s_j - s_i) \geq 0$ for all $\alpha_1, \ldots, \alpha_n \in \mathbf{C}$ and $s_1, \ldots, s_n \in G$. We call f **positive definite.**

2.2 Theorem: *A continuous positive definite function on G is weakly almost periodic.*

Proof: Let $f \in C^B(G)$ be positive definite. There exists a continuous unitary cyclic representation $t \mapsto U_t$ of the group G such that $f(t) = (U_t h_0, h_0)$ where h_0 is the cyclic vector in the Hilbert space $\mathscr{H}$ [N, p. 393; see also 7.1.5]. Let $\{s_n\}$ be a sequence in G.

Define the linear transformation $T: \mathscr{H} \to C^B(G)$ by $T(k) = \phi$ where $\phi(t) = (U_t h_0, k)$. Now

$$\| T(k) \|_\infty = \| \phi \|_\infty \leq \| U_t h_0 \| \; \| k \| = \| h_0 \| \; \| k \| .$$

Thus $\| T \| \leq \| h_0 \|$. So T is strongly continuous and hence weakly continuous [DS, p. 422]. Now

$$f_{s_n}(t) = (U_{s_n + t} h_0, h_0) = (U_t h_0, U_{s_n}^{-1} h_0) = T(U_{s_n}^{-1} h_0).$$

Also $\| U_{s_n}^{-1} h_0 \| = \| h_0 \|$ for all n, and the unit sphere in $\mathscr{H}$ is weakly compact and so weakly sequentially compact. Thus there exists a subsequence $\{s_i\}$ and $h' \in \mathscr{H}$ such that $U_{s_i}^{-1} h_0 \xrightarrow{i} h'$ weakly. Let $f' \in C^B(G)$ be such that $f' = T(h')$. Now the weak continuity of T implies that $\{f_{s_i}\}$ converges weakly to f'. Thus f is weakly almost periodic. □

2.3 Theorem: *$WAP(G)$ is a translation invariant B^*-subalgebra of $C^B(G)$.*

Proof: The linearity and invariance is direct. Let $\{f_n\} \subset WAP(G)$ be such that $f_n \to f$ uniformly. Let $\{s_i\}$ be a sequence in G. We extract from $\{s_i\}$ via a diagonal process a subsequence $\{s_j\}$ such that $(f_n)_{s_j}$ has a weak limit g_n for each n. Now

$$\begin{aligned}\| g_n - g_m \|_\infty &= \sup \{ | \textstyle\int (g_n - g_m)\, d\mu | : \mu \in M(G), \| \mu \| = 1\} \\ &= \sup \{\lim_j | \textstyle\int ((f_n)_{s_j} - (f_m)_{s_j})\, d\mu | : \mu \in M(G), \| \mu \| = 1\} \\ &\leq \| f_n - f_m \|_\infty .\end{aligned}$$

Thus $\{g_n\}$ is a Cauchy sequence in $C^B(G)$ with strong (hence weak) limit g. It follows that g is a weak cluster point of $\{f_{s_j}\}$. Thus the orbit of f is weakly countably compact and so f is w.a.p. Hence we have that $WAP(G)$ is a closed linear subspace of $C^B(G)$. We show now that $WAP(G)$ is closed under pointwise multiplication. Let $f, g \in WAP(G)$. Pick $\{s_n\} \subset G$. Choose a subsequence $\{s_i\} \subset \{s_n\}$ and $f', g' \in C^B(G)$ such that $f_{s_i} \xrightarrow{i} f'$ and $g_{s_i} \xrightarrow{i} g'$ weakly. View $C^B(G)$ as the continuous functions on the Stone–Čech compactification, βG. Thus $(f \cdot g)_{s_i} \xrightarrow{i} f' \cdot g'$ pointwise on βG and hence weakly in $C^B(G)$. Finally note that if $f \in WAP(G)$, then so is $\bar{f}$. Thus $WAP(G)$ is a B^*-subalgebra of $C^B(G)$. □

2.4 Corollary: *$M(\Gamma)\hat{\ }^{-} \subset WAP(G)$. In particular, $C_0(G) \subset WAP(G)$.*

2.5 Theorem: *A weakly almost periodic function is uniformly continuous.*

Proof: Let $f \in WAP(G)$. Suppose f is not uniformly continuous. Thus there is $\varepsilon > 0$ such that for any symmetric neighborhood, V, about 0 in G we can find $x, y \in G$ such that $x - y \in V$ but $| f(x) - f(y) | \geq \varepsilon$. Thus we can construct a net $(x_\alpha, y_\alpha) \in G \times G$ such that $x_\alpha - y_\alpha \xrightarrow{\alpha} 0$ and $| f(x_\alpha) - f(y_\alpha) | \geq \varepsilon$. Let $g \in WAP(G)$ be defined by $g_\alpha(t) = f(t + x_\alpha) - f(t + y_\alpha)$. Let h be a weak cluster point of $\{g_\alpha\}$ (h exists since f is w.a.p.). Now $| g_\alpha(0) | \geq \varepsilon$. We arrive at the required contradiction by showing $h = 0$.

If suffices to show that $\int_K h(t)\, dt = 0$ for every compact set K. We show this by noting that $\int_K g_\alpha(t)\, dt \xrightarrow{\alpha} 0$. Now

$$\begin{aligned}| \textstyle\int_K g_\alpha(t)\, dt | &= | \textstyle\int_K (f(t + x_\alpha) - f(t + y_\alpha))\, dt | \\ &\leq | \textstyle\int_{K - x_\alpha} f(t)\, dt - \int_{K - y_\alpha} f(t)\, dt | \leq \| f \|_\infty m(\Delta_\alpha)\end{aligned}$$

where $m(\Delta_\alpha)$ is the Haar measure of the symmetric difference of $K - x_\alpha$ and $K - y_\alpha$. Now $m(\Delta_\alpha) \xrightarrow{\alpha} 0$ since $x_\alpha - y_\alpha \xrightarrow{\alpha} 0$ [H, p. 266]. □

2.6 Remark: For $f \in C^B(G)$, $O(f)$ denotes the orbit of f; that is, the set of translates of f. Let $co(O(f))$ denote the convex hull of $O(f)$. Using ergodic considerations, Eberlein has shown that for each $f \in WAP(G)$, $\overline{co(O(f))}$ contains a unique constant function [1, p. 225]. Furthermore, this constant coincides with the von Neumann mean value [1, p. 236]. Also every $f \in WAP(G)$ has a decomposition as $f = f_a + f_0$ where f_a is almost periodic (that is, the orbit of f is relatively strongly compact) and $f_0 \in WAP(G)$ with the mean value of $| f_0 |$ zero [2, p. 138].

3. Relationship to Fourier-Stieltjes Transforms

3.1 Definition: Let G be discrete. Let $E \subset G$ be such that $(E + y_1) \cap (E + y_2)$ is finite for all $y_1, y_2 \in G$ with $y_1 \neq y_2$. We call E a T-set.

3.2 Theorem: *Let G be discrete and $E \subset G$ a T-set. Then if $f \in C^B(G)$ has* spt$(f) = E$, *then $f \in WAP(G)$.*

Proof: A net of functions, $\{f_\alpha\}_{\alpha \in A}$, in $C^B(G)$ is said to converge **quasi-uniformly** to f on G if $f_\alpha \to f$ pointwise on G and for all $\varepsilon > 0$ and $\alpha_0 \in A$, there exist $\alpha_1, \ldots, \alpha_k \geq \alpha_0$ such that for each $x \in G$, $\min_{1 \leq i \leq k} | f_{\alpha_i}(x) - f(x) | < \varepsilon$. A bounded sequence, $\{f_n\}$, in $C^B(G)$ converges weakly to f if and only if $\{f_n\}$ and every subsequence of $\{f_n\}$ converges to f quasi-uniformly on G [DS, p. 281].

Let $\{x_i\} \subset G$ be a sequence. We may assume that $x_i \neq x_j$ for $i \neq j$. We write f_i for f_{x_i}. Let $N \in Z_+$ and $\varepsilon > 0$. Suppose there exist $p, q \in G$ such that $f_i(p) \neq 0 \neq f_i(q)$ for infinitely many i. Then $(\text{spt}(f) - p) \cap (\text{spt}(f) - q)$ is infinite and so $p = q$.

Case 1: Suppose $\lim_i f_i(x) = 0$ for all $x \in G$:

Now $\text{spt}(f_N) \cap \text{spt}(f_{N+1})$ is finite, say $\{y_3, \ldots, y_m\}$. By our case assumption we can find $i_3, \ldots, i_m > N$ such that $| f_{i_j}(y_j) | < \varepsilon$, $3 \leq j \leq m$. Let $i_1 = N$ and $i_2 = N + 1$. Then $\min\{ | f_{i_j}(x) | : 1 \leq j \leq m\} < \varepsilon$ for all $x \in G$. Any subsequence of $\{f_i\}$ has the same properties. Thus $f_i \xrightarrow{i} 0$ weakly.

Case 2: Suppose there exists a unique $y \in G$ such that $\lim_i \sup | f_i(y) | \neq 0$:

Taking a subsequence, we may assume that $\lim_i f_i(y)$ exists; call it c. As in Case 1, we find $i_1, \ldots, i_m \geq N$ such that

$$\min \{ | f_{i_j}(x) - g(x) | : 1 \leq j \leq m\} < \varepsilon$$

for all $x \in G$ where $g(t) = 0$ if $t \neq y$ and $g(y) = c$. Any subsequence of $\{f_i\}$ has the same properties so $f_i \xrightarrow{i} g$ weakly. We will use this method in Theorem 3.6. □

3.3 Theorem: *Let G be a discrete group with Z as a subgroup. Let*

$$E = \{nk! : 1 \leq n \leq k, k = 1, 2, \ldots\}.$$

Let χ_E denote the characteristic function of E. Then $\chi_E \in WAP(G)$ but $\chi_E \notin \overline{M(\Gamma)\hat{}}$, where Γ is the (compact) character group of G.

Proof: To show that $\chi_E \in WAP(G)$ we show that E is a T-set. Assume $n > 0$ and let $t \in E \cap (E + n)$. Then $t = lm! = pq! + n$ where $1 \leq p \leq q, 1 \leq l \leq m$. Thus $q \leq m$.

CASE 1: Suppose $q < m$:

Then $n = lm! - pq! \geq m! - (m-1)(m-1)! \geq m - 1$. So $m \leq n + 1$. Thus $t = lm! \leq mm! \leq (n+1)(n+1)!$. So in this case, $E \cap (E + n)$ is finite.

CASE 2: Suppose $q = m$:

Since $n > 0$, $l > p$. Then

$$n = lm! - pq! = lm! - pm! = (l-p)m! \geq m! \geq m.$$

So $m \leq n$. As before this implies that $E \cap (E + n)$ is finite. Thus $\chi_E \in WAP(G)$.

We now show that $\chi_E \notin M(\Gamma)\hat{}^{-}$. Let $M > 0$ be such that

$$\left| \sum_{n=1}^{k} \frac{\sin(nt)}{n} \right| < M \text{ for all } k, \text{ all } t$$

[RP, p. 180]. Let

$$S_k = \sum_{n=1}^{k} \frac{1}{n}$$

and

$$\lambda_k = \frac{1}{2S_k}\left(-\frac{1}{k}\delta_{-kk!} - \cdots - \delta_{k!} + \delta_{k!} + \frac{1}{2}\delta_{2k!} + \cdots + \frac{1}{k}\delta_{kk!}\right)$$

where δ_p denotes the point mass at p. Thus $\| \lambda_k \| = 1$ and for $\gamma \in \Gamma$ with $\gamma(k!) = e^{i\theta}$, we have

$$|\hat{\lambda}_k(\gamma)| = \frac{1}{2S_k}\left| \sum_{n=1}^{k} \frac{1}{n}(e^{in\theta} - e^{-in\theta}) \right| = \frac{1}{2S_k}\left| 2i \sum_{n=1}^{k} \frac{\sin(n\theta)}{n} \right|$$

$$\leq \frac{M}{S_k} \xrightarrow{k} 0.$$

By the characterization theorem for $M(\Gamma)\hat{}^{-}$ (see 3.3.12), $\chi_E \in M(\Gamma)\hat{}^{-}$ would imply that $\Sigma_\Gamma\, \chi_E\, d\lambda_k \xrightarrow{k} 0$. But $\Sigma_\Gamma\, \chi_E\, d\lambda_k = \frac{1}{2}$. So $\chi_E \notin M(\Gamma)\hat{}^{-}$. This proof has a more general setting as we will see in Theorems 3.5 and 3.6. □.

3.4 Lemma: *Let G be an abelian group which is not of bounded order. Then G contains a set E such that* (a) *E is the union of disjoint sets A_k ($k = 1, 2, 3, \ldots$) each of which consists of k distinct elements $x_k, 2x_k, \ldots, kx_k$.* (b) *If $x \in G$ and $x \neq 0$, then $E \cap (E + x)$ is a finite set.* (c) $A_k \cap (-A_k) = \emptyset$.

Proof:

CASE 1: Suppose G contains an element x of infinite order:

Then the sets $A_k = \{nk!\,x,\ 1 \leq n \leq k\}$ will do as in Theorem 3.3.

CASE 2: Suppose that G contains no element of infinite order:

Let x_1 be any element in G such that $x_1 \neq -x_1$. Let $A_1 = \{x_1\}$. Suppose $A_1, \ldots, A_{k-1} (k \geq 2)$ have been constructed. Let H_{k-1} be the subgroup generated by $A_1, \ldots, A_{k-1}$. By our case assumption, we are able to find $x_k \in G$ such that $\{x_k, \ldots, 2kx_k\} \cap H_{k-1}$ is empty. Let $A_k = \{x_k, \ldots, kx_k\}$

and $E = A_1 \cup A_2 \ldots$. Conditions (a) and (c) are now immediate. To show (b), we first note that $E \cap (E + x)$ is empty if x is in no H_k. Suppose then that x is in some H_k. Let k denote the smallest integer such that $x \in H_k$. It suffices to show that $E \cap (E + x) \subset H_k$. Let $y \in E \cap (E + x)$ and $y \notin H_k$. Then $y = x + z$, with $z \in E$. There exist n, m so that $y \in A_n$, $z \in A_m$. Now since $y \notin H_k \supset A_k$, $k < n$. Since $y - x \in H_n$, $z = y - x \in H_n$ so $m \leq n$. If $m = n$, then the construction of A_i's implies that $y - z \notin H_k$ since $k < n$. But $y - z = x \in H_k$. This is the required contradiction. If $m < n$, then $x + z \in H_p$ where $p = \max(m, k)$ so that $y \in H_p$. Since $p < n$, the construction of A_i's implies that $x + z = y \notin A_n$. But $y \in A_n$. This is the required contradiction. □

3.5 Theorem: *Let G be an infinite discrete abelian group. Then $M(\Gamma)\hat{}^{-} \neq WAP(G)$, where Γ is the (compact) character group of G.*

Proof:

CASE 1: Suppose G is not of bounded order:

Using the previous lemma, one repeats the proof of Theorem 3.3.

CASE 2: Suppose G is of bounded order:

Let $Z(p)$ denote the finite cyclic group of p elements of unimodular complex numbers, and $Z(p)^\infty$ the weak direct product of $Z(p)$ over a countable infinite index set. Thus there exists p such that $Z(p)^\infty$ is a subgroup of G. We denote $Z(p)^\infty \subset G$ by H.

There exists $\lambda_n \in M(H) \subset M(G)$ such that $\| \lambda_n \| = 1$ and $\| \hat{\lambda}_n \|_\infty \leq 1/n$ $(n = 1, 2, \ldots)$ since $M(H)\hat{}$ is not complete in the sup-norm topology for G infinite (see 3.2.2). Let S_n denote the spt (λ_n). We may assume that the S_n's are finite sets and pairwise disjoint. Let f be a (continuous) bounded function on G such that $\Sigma_G\, f\, d\lambda_n = \| \lambda_n \| = 1$, $\| f \|_\infty \leq 1$, and $\operatorname{spt}(f) = \bigcup_{n=1}^\infty S_n$. By the characterization theorem for $M(\Gamma)\hat{}^{-}$ (see 3.3.12) if $f \in M(\Gamma)\hat{}^{-}$, then $\Sigma_G\, f\, d\lambda_n \xrightarrow{n} 0$. But $\Sigma_G\, f\, d\lambda_n = 1$ so $f \notin M(\Gamma)\hat{}^{-}$. It remains to show that we may pick f such that $f \in WAP(G)$.

Let $S = \bigcup_{n=1}^\infty S_n$. We will construct S such that $S \cap (S + y)$ is finite for every $y \neq 0$ in G. Thus S will be a T-set and by Theorem 3.2, any bounded function supported on a T-set is w.a.p. Let $F \subset H$ be a finite set. Let $\alpha(F)$ $(\beta(F))$ denote the last (first) coordinate such that all elements of F are 1 for coordinates $< \alpha(F)$ $(> \beta(F))$. Let the S_n's be constructed such that $\alpha(S_1) < \beta(S_1) < \alpha(S_2) < \beta(S_2) < \cdots < \alpha(S_n) < \beta(S_n)$. Thus $S = \bigcup_{n=1}^\infty S_n$ is a T-set and $f \in WAP(G) \backslash M(\Gamma)\hat{}^{-}$. □

3.6 Theorem: *Let G be a locally compact group which contains a copy of $\mathbf{R}^n$, $n \geq 1$. Then $WAP(G) \neq M(\Gamma)\hat{}^{-}$.*

Proof: We may assume that the real line, $\mathbf{R}$, is a direct summand. It will suffice to show that $M(\mathbf{R})\hat{}^{-} \neq WAP(\mathbf{R})$.

Let $E = \{nk!: 1 \leq n \leq k,\ k = 1, 2, \ldots\}$. Let f be defined on $\mathbf{R}$ by

$f(x) = 1 - 8|y - x|$ for $x \in (y - \frac{1}{8}, y + \frac{1}{8})$ where $y \in E$ and $f(x) = 0$ otherwise. It follows as in Theorem 3.3, that $f \notin M(\mathbf{R})\hat{}^{-}$. It remains to show that $f \in WAP(\mathbf{R})$. Let $\{s_n\}$ be a sequence of real numbers. Suppose $\{s_n\}$ has a finite cluster point, s. The uniform continuity of f implies that f_s is a weak cluster point of $\{f_{s_n}\}$. Thus $O(f)$ is weakly countably compact and thus relatively weakly compact. If $\{s_n\}$ has no finite cluster point, then we pick a subsequence $\{s_i\}$ such that $|s_i - s_j| > 1$ for $i \neq j$.

CLAIM 1: If $i \neq j$, then $(\mathrm{spt}\,(f) - s_i) \cap (\mathrm{spt}\,(f) - s_j)$ is compact:

Let S be the $\mathrm{spt}(f)$ and let $x \in (S - s_i) \cap (S - s_j)$. It suffices to show that $(S - s_i) \cap (S - s_j)$ is bounded. Write $x = t + \delta - s_i$ where $|\delta| \leq \frac{1}{8}$ and $t \in E$, and $x = t' + \delta' - s_j$ where $|\delta'| \leq \frac{1}{8}$ and $t' \in E$. Now

$$t' - t = s_j - s_i + \delta - \delta' \in s_j - s_i + [-\tfrac{1}{4}, \tfrac{1}{4}].$$

Let y be the unique integer in $s_j - s_i + [-\frac{1}{4}, \frac{1}{4}]$ so $t' - t = y$. Now $y \neq 0$ since $|s_j - s_i| > 1$. Now $x - \delta' + s_j = t' = y + t$ so

$$x \in (E \cap (y + E)) + [-\tfrac{1}{8}, \tfrac{1}{8}] - s_j$$

which is a bounded set.

CLAIM 2: If $f(x_o + s_i) \neq 0 \neq f(x + s_i)$ for i in an infinite set I, then $|x - x_o| \leq \frac{1}{4}$:

For $i \in I$, let $x_o = t_i + \delta - s_i$ and $x = t_i' + \delta' - s_i$ where $t_i, t_i' \in E$ and $|\delta|, |\delta'| \leq \frac{1}{8}$. Hence $t_i' - t_i \in x - x_o + [-\frac{1}{4}, \frac{1}{4}]$. Let y be the unique integer in $x - x_o + [-\frac{1}{4}, \frac{1}{4}]$. Thus $t_i' - t_i = y$, for all $i \in I$. Now $|s_i - s_j| > 1$ for $i \neq j$ implies that $t_i \neq t_j$ for $i \neq j$. Hence $E \cap (E - y)$ contains each t_i, $i \in I$, and so is infinite. Thus $y = 0$ and $|x - x_o| \leq \frac{1}{4}$.

Now if there is a subsequence of $\{f_{s_i}\}$, say $\{f_{s_j}\}$, which converges to 0 at each point of G, then Claim 1 implies as in Theorem 3.2 that $f_{s_j} \xrightarrow{j} 0$ weakly by quasi-uniform convergence. Otherwise, there is $x_o \in G$ and a subsequence, $\{f_{s_j}\}$, of $\{f_{s_i}\}$ such that $f_{s_j} \xrightarrow{j} 0$ pointwise outside of $[x_o - \frac{1}{4}, x_o + \frac{1}{4}]$ by Claim 2. Since f is uniformly continuous, $\{f_{s_j}\}$ is an equicontinuous family. Hence there is a subsequence, $\{f_{s_k}\}$, such that $\{f_{s_k}\}$ converges uniformly in $[x_o - \frac{1}{4}, x_o + \frac{1}{4}]$. Once again, $\{f_{s_k}\}$ converges weakly by quasi-uniform convergence. □

3.7 Theorem: *Let G be any infinite noncompact locally compact group. Then $WAP(G) \neq M(\Gamma)\hat{}^{-}$.*

Proof: We first note that if G is compact, then

$$M(\Gamma)\hat{}^{-} = WAP(G) = C^B(G).$$

If G contains a copy of $\mathbf{R}^n$, then Theorem 3.6 applies. Otherwise, the structure theorem for locally compact abelian groups [HR, p. 389] implies that G contains a compact open subgroup, Λ. Now G/Λ is infinite and discrete. Apply Theorem 3.5 to find f on G/Λ which is w.a.p. on G/Λ but not in $M((G/\Lambda)\hat{})\hat{}^-$. Extend f to f' on G by $f'(x) = f(x + \Lambda)$, $x \in G$. Now $f' \in WAP(G)$ by quasi-uniform convergence.

Since $f \notin M((G/\Lambda)\hat{})\hat{}^-$, there are $\lambda_n \in M(G/\Lambda)$, $\| \lambda_n \| = 1$, $\hat{\lambda}_n \to 0$ pointwise and $\Sigma_{G/\Lambda} f \, d\lambda_n \overset{n}{\nrightarrow} 0$. We may assume that spt (λ_n) is a finite set. Let $\lambda_n = c_{n,1} \delta_{\bar{x}_{n,1}} + \ldots$ (m_n terms) where $\bar{x}_{n,k} = x_{n,k} + \Lambda$ in G/Λ. Extend λ_n to μ_n on G by smearing points to cosets. Thus $\mu_n = c_{n,1} \delta_{n,1} + \cdots + c_{n,m_n} \delta_{n,m_n}$ where $\delta_{n,k}$ is Haar measure, m, on G restricted to $x_{n,k} + \Lambda$. We may assume that $m(\Lambda) = 1$. Now for $\gamma \in G\hat{}$,

$$\hat{\mu}_n(\gamma) = (c_{n,1}\, \gamma(x_{n,1}) + \cdots + c_{n,m_n}\, \gamma(x_{n,m_n})) \int_\Lambda \gamma(x)\, dm.$$

If $\gamma \not\equiv 1$ on Λ, then $\int_\Lambda \gamma(x)\, dm = 0$. If $\gamma \equiv 1$ on Λ, then γ may be considered as an element, γ^*, of $(G/\Lambda)\hat{}$ and $\hat{\mu}_n(\gamma) = \hat{\lambda}_n(\gamma^*)$. In particular, $\hat{\mu}_n(\gamma) \overset{n}{\to} 0$ for all $\gamma \in G\hat{}$. In summary, $\{\mu_n\} \subset M(G)$, $\| \mu_n \| = 1$, $\hat{\mu}_n \overset{n}{\to} 0$ pointwise, and $\int_G f' \, d\mu_n = \Sigma_{G/\Lambda} f \, d\lambda_n \overset{n}{\nrightarrow} 0$. Thus $f' \in WAP(G) \backslash M(\Gamma)\hat{}^-$. □

4. Historical Notes

4.1: The basic theory of w.a.p. functions (Theorems 2.2, 2.3, 2.4) was developed by Eberlein [1, 2]. Rudin [2] was the first to show that $M(\Gamma)\hat{}^-$ could be proper in $WAP(G)$. He showed this to be the case provided G contained a closed discrete subgroup which is not of bounded order. We have modified his proof using a technique from Ramirez [2]. This avoids a difficult trigonometric inequality. We also have introduced quasi-uniform convergence. This avoids going outside the group to its Stone-Čech compactification. The general result is given in Ramirez [3].

The weakly almost periodic functions have been studied extensively on semigroups, see deLeeuw and Glicksberg [1, 2, 3], Pym [1], and Berglund and Hofmann [BH].

CHAPTER 5

SIDON SETS

1. Introduction

1.1: In this chapter, G will denote a **compact** abelian group with a (discrete) dual group Γ. We characterize Sidon sets, and then give sufficient conditions for a set to be a Sidon set.

1.2 Definition: Let E be a subset of Γ. A function in $L^1(G)$ is called an E-function if $\hat{f}(\gamma) = 0$ for all γ not in E. We call f bounded if $\| f \|_\infty < \infty$.

A trigonometric polynomial which is an E-function will be called an E-polynomial.

A set E is called a **Sidon set** if and only if there is a finite constant B such that

$$\sum_{\gamma \in \Gamma} |\hat{f}(\gamma)| \leq B \| f \|_\infty$$

for every E-polynomial f.

We now state the fundamental characterization of Sidon sets. A proof can be found in Rudin's book [R, p. 121]. The proof is based on the closed graph theorem. We then give the result that the interpolation problem is equivalent to the approximate interpolation problem for $M(G)\hat{}$, [R, p. 123].

1.3 Theorem: *Let $E \subset \Gamma$. The following are equivalent:*

(A) *E is a Sidon set (with constant B).*

(B) *Every bounded E-function f has $\Sigma_{\gamma \in \Gamma} | \hat{f}(\gamma) | \leq B \| f \|_\infty$.*

(C) *Every continuous E-function f has $\Sigma_{\gamma \in \Gamma} | \hat{f}(\gamma) | < \infty$.*

(D) *If* $\phi \in L^\infty(E)$, *then there is a* $\mu \in M(G)$ *such that* $\hat{\mu}(\gamma) = \phi(\gamma)$ *for all* $\gamma \in E$ *and* $\|\mu\| \le B\|\phi\|_\infty$.

(E) $L^\infty(E) = M(G)\hat{}\,|\,E$.

(F) *If* $\phi \in C_0(E)$, *then there is an* $f \in L^1(G)$ *such that* $\hat{f}(\gamma) = \phi(\gamma)$ *for all* $\gamma \in E$ *and* $\|f\|_1 \le B\|\phi\|_\infty$.

(G) $C_0(E) = L^1(G)\hat{}\,|\,E$.

2. Other Equivalences

2.1 Theorem: *Let* $E \subset \Gamma$. *The following are equivalent:*

(A) *E is a Sidon set.*

(H) *For every function* ϕ *on E with* $\phi(\gamma) = \pm 1$, *there corresponds a measure* $\mu \in M(G)$ *such that*

$$\sup_{\gamma \in E} |\hat{\mu}(\gamma) - \phi(\gamma)| < 1.$$

Proof: If E is a Sidon set, then (H) follows from property (E). Thus (A) implies (H).

Now suppose (H) holds. We show property (C). We may assume that f is a continuous E-function with $\hat{f}$ real. Define ϕ on Γ so that $\phi = \pm 1$ and $\phi \hat{f} = |\hat{f}|$. Let $\mu \in M(G)$ be such that

$$\sup_{\gamma \in E} |\hat{\mu}(\gamma) - \phi(\gamma)| \le 1 - \delta, \text{ for some } \delta,\ 0 < \delta \le 1.$$

Let $\sigma = \frac{1}{2}(\mu + \mu^*)$. Then $\hat{\sigma} = \operatorname{Re}(\hat{\mu})$ so

$$\sup_{\gamma \in E} |\hat{\sigma}(\gamma) - \phi(\gamma)| \le 1 - \delta.$$

Also

$$|\hat{f}\hat{\sigma} - |\hat{f}|| = |\hat{f}|\,|\hat{\sigma} - \phi| \le |\hat{f}|(1 - \delta).$$

Thus $\delta|\hat{f}| \le \hat{f}\hat{\sigma}$. Set $g = f * \sigma$. So $\hat{g} \ge \delta|\hat{f}|$. Let $\gamma_1, \ldots, \gamma_n \in \Gamma$, and let k be a trigonometric polynomial on G such that $\|k\|_1 < 2$, $\hat{k} \ge 0$, and $\hat{k}(\gamma_i) = 1$ $(1 \le i \le n)$. Now $k * g$ is a trigonometric polynomial, and

$$\delta \sum_{i}^{n} |\hat{f}(\gamma_i)| \le \sum_{i=1}^{n} \hat{f}(\gamma_i)\hat{\sigma}(\gamma_i) = \sum_{i=1}^{n} \hat{k}(\gamma_i)\,\hat{g}(\gamma_i)$$

$$\le \sum_{\gamma \in \Gamma} \hat{k}(\gamma)\,\hat{g}(\gamma) \le \|k\|_1\,\|g\|_\infty \quad \text{(by Parseval's formula)}$$

$$\le 2\,\|f\|_\infty\,\|\sigma\|.$$

Hence $\Sigma_{\gamma \in \Gamma} |\hat{f}(\gamma)| < \infty$. Thus E is a Sidon set. $\square$

2.2 Corollary: *Let $E \subset \Gamma$. The following are equivalent:*

(A) *E is a Sidon set.*

(I) *$M(G)\hat{}^{-} \mid E = L^\infty(E)$ (closure in sup-norm on Γ).*

(J) *$\overline{M(G)\hat{} \mid E} = L^\infty(E)$ (closure in sup-norm on E).*

2.3 Theorem: *Let $E \subset \Gamma$. The following are equivalent:*

(A) *E is a Sidon set.*

(K) *For all $f \in L^\infty(E)$, if $\{\lambda_n\} \subset M(E)$, $\| \hat{\lambda}_n \|_\infty \le 1$, and $\hat{\lambda}_n(x) \xrightarrow{n} 0$ for all $x \in G$, then $\Sigma_{\gamma\in\Gamma} f(\gamma)\, d\lambda_n(\gamma) \xrightarrow{n} 0$.*

(L) *For all $f \in L^\infty(E)$, if $\{\lambda_n\} \subset M(E)$, $\| \lambda_n \| \le 1$, and $\hat{\lambda}_n(x) \xrightarrow{n} 0$ for all $x \in G$, then $\Sigma_{\gamma\in\Gamma} f(\gamma)\, d\lambda_n(\gamma) \xrightarrow{n} 0$.*

(M) *For all $f \in L^\infty(E)$, if $\{\lambda_n\} \subset M(E)$, $\| \lambda_n \| \le 1$, and $\| \hat{\lambda}_n \|_\infty \xrightarrow{n} 0$, then $\Sigma_{\gamma\in\Gamma} f(\gamma)\, d\lambda_n(\gamma) \xrightarrow{n} 0$.*

Proof: Let E be a Sidon set and $f \in L^\infty(E)$. There is $\mu \in M(G)$ such that $\hat{\mu} \mid E = f$. Hence

$$\sum_{\gamma\in E} f\, d\lambda = \sum_{\gamma\in E} \hat{\mu}\, d\lambda = \sum_{\gamma\in\Gamma} \hat{\mu}\, d\lambda = \int_G \hat{\lambda}\, d\mu,$$

for $\lambda \in M(E)$. The Lebesgue dominated convergence theorem now yields that (A) implies (K).

Clearly, (K) implies (L) and (L) implies (M).

Assume (M) is satisfied. If E is not a Sidon set, we may find $\lambda_2 \in M(E)$, $\| \lambda_2 \| = 1$, and $\| \hat{\lambda}_2 \|_\infty \le \frac{1}{2}$. We may assume that $F_2 = \operatorname{spt}(\lambda_2)$ is finite. Since F_2 is finite and E is not a Sidon set, then $E \backslash F_2$ is not a Sidon set. Similarly there exists $\lambda_3 \in M(E\backslash F_2)$, $\| \lambda_3 \| = 1$, $F_3 = \operatorname{spt}(\lambda_3)$ finite, and $\| \hat{\lambda}_3 \|_\infty \le 1/3$. Likewise, there exists

$$\lambda_{n+1} \in M(E\backslash F_2 \cup F_3 \cup \cdots \cup F_n),$$

$$\| \lambda_{n+1} \| = 1,\ F_{n+1} = \operatorname{spt}(\lambda_{n+1}) \text{ finite, and } \| \hat{\lambda}_{n+1} \|_\infty \le 1/(n+1).$$

Let $f \in L^\infty(E)$, $\| f \|_\infty \le 1$ be such that $\Sigma_{\gamma\in F_n} f\, d\lambda_n = 1$. Now $\{\lambda_n\} \subset M(E)$, $\| \lambda_n \| = 1$, $\| \hat{\lambda}_n \|_\infty \xrightarrow{n} 0$, but $\Sigma_{\gamma\in E} f\, d\lambda_n = 1$. This is the required contradiction. Thus E is a Sidon set. □

2.4 Corollary: *Let $E = \{\gamma_k\}_{k=1}^\infty \subset \Gamma$. The following are equivalent:*

(A) *E is a Sidon set.*

(N) *Suppose that there is an infinite set, S, of positive integers $0 < N_1 < N_2 < \cdots$ such that for any sequence of complex numbers $\{c_k\}_{k=1}^\infty$ if*

$$\sup_{N_j\in S} \left| \sum_{k=1}^{N_j} c_k \gamma_k(x) \right| \le M_x < \infty$$

for all $x \in G$, then $\sum_{k=1}^{\infty} |c_k| < \infty$.

Proof: Let E be a Sidon set. Suppose that there is a sequence $\{c_k\}_{k=1}^{\infty}$ such that

$$\sup_{N_j \in S} \left| \sum_{k=1}^{N_j} c_k \gamma_k(x) \right| \le M_x$$

but

$$\sum_{k=1}^{\infty} |c_k| = \infty.$$

Let $\lambda_j \in M(Z)$ $(N_j \in S)$ be defined by

$$\lambda_j = \frac{\sum_{k=1}^{N_j} c_k \delta_{\gamma_k}}{1 + \sum_{k=1}^{N_j} |c_k|}$$

where δ_{γ_k} denotes point mass at γ_k. Now $\| \lambda_j \| \le 1$ and

$$|\hat{\lambda_j}(x)| = \frac{\left| \sum_{k=1}^{N_j} c_k \gamma_k(x) \right|}{1 + \sum_{k=1}^{N_j} |c_k|} \le \frac{M_x}{1 + \sum_{k=1}^{N_j} |c_k|} \xrightarrow{j} 0.$$

Let f be defined on E by $f(\gamma_k) \cdot c_k = |c_k|$. By the above theorem, since E is a Sidon set, (L) holds. But

$$\sum_{\gamma \in \Gamma} f(\gamma)\, d\lambda_j(\gamma) = \frac{\sum_{k=1}^{N_j} f(\gamma_k) \cdot c_k}{1 + \sum_{k=1}^{N_j} |c_k|} = \frac{\sum_{k=1}^{N_j} |c_k|}{1 + \sum_{k=1}^{N_j} |c_k|} \xrightarrow{j} 1.$$

This is a contradiction and so $\Sigma_{k=1}^{\infty} |c_k| < \infty$ and (A) implies (N).

Now suppose that condition (N) holds. If E is not a Sidon set, then as in the previous theorem, we may find $\{\lambda_j\} \subset M(E)$ such that $\text{spt}(\lambda_j)$ are finite and pair-wise disjoint, $\| \lambda_j \| = 1$, $\| \hat{\lambda_j} \|_{\infty} \le (\frac{1}{2})^j$. Let

$$\lambda_1 = \sum_{k=1}^{N_i} c_k \delta_{\gamma_k}, \text{ and } \lambda_j = \sum_{k=N_{j-1}+1}^{N_j} c_k \delta_{\gamma_k}, j > 1.$$

Now $\Sigma_{k=1}^{N_j} c_k \gamma_k(x)$ is (pointwise) bounded for all j, but $\Sigma_{k=1}^{\infty} |c_k| = \infty$. This is the required contradiction. Hence (N) implies (A). □

2.5 Theorem: *Let $E \subset \Gamma$. The following are equivalent:*

(A) *E is a Sidon set.*

(O) *If E_1 and E_2 are disjoint subsets of E, then $\overline{E}_1$ and $\overline{E}_2$ are disjoint subsets in $\overline{\Gamma}$ (closure in the maximal ideal space of $M(G)$).*

Proof: Suppose E is a Sidon set. Let E_1 and E_2 be disjoint subsets of E. Let

$\phi \in L^{\infty}(E) = 1$ on E_1 and 2 on E_2. Thus there is $\mu \in M(G)$ such that $\hat{\mu} = 1$ on E_1 and 2 on E_2. Thus E_1 and E_2 have disjoint closures in the maximal ideal space of $M(G)$. Hence (A) implies (O).

Now assume (O). Let $\phi \in L^{\infty}(E)$ be such that $\phi(\gamma) = \pm 1$. Let $E_1 = \{\gamma \in E: \phi(\gamma) = 1\}$ and $E_2 = \{\gamma \in E: \phi(\gamma) = -1\}$. Thus E_1 and E_2 are disjoint. Since Γ is contained in the symmetric maximal ideals, $M(G)^{\tilde{}} \mid \overline{\Gamma}$ is dense in $C(\overline{\Gamma})$ by the Stone-Weierstrass theorem. Since $\overline{E}_1$ and $\overline{E}_2$ are disjoint, there is $f \in C(\overline{\Gamma})$ such that $f = 1$ on E_1 and -1 on E_2. Let $\mu \in M(G)$ with $\| \tilde{\mu} \mid \overline{\Gamma} - f \|_{\infty} < \frac{1}{2}$. Thus (H) is satisfied with $\delta = \frac{1}{2}$. Hence E is a Sidon set and (O) implies (A). □

2.6 Corollary: *Let E and F be Sidon sets in* Γ. *If* $\chi_E \in M(G)^{\hat{}-}$ *(closure in sup-norm on* Γ*), then* $E \cup F$ *is a Sidon set* (see Section 5).

Proof: We may assume that E and F are disjoint.

Let $H = E \cup F$ and H_1, H_2 be disjoint subsets of H. Let $E_i = H_i \cap E$ and $F_i = H_i \cap F (i = 1, 2)$. Since $\chi_E \in M(G)^{\hat{}-}$, $\overline{E}$ and $\overline{F}$ are disjoint (closure in the maximal ideal space of $M(G)$). It follows that $\overline{H}_1 = \overline{E_1 \cup F_1} = \overline{E}_1 \cup \overline{F}_1$ and $\overline{H}_2 = \overline{E_2 \cup F_2} = \overline{E}_2 \cup \overline{F}_2$ are disjoint. Hence $E \cup F$ is a Sidon set by Theorem 2.5. □

2.7 Remark: Theorem 1.3 gives the equivalence between Sidon sets and the absolute convergence of Fourier series. Theorem 2.4 gives the equivalence between Sidon sets and the absolute convergence of trigonometric series. Theorem 2.1 gives the result that the interpolation problem is equivalent to the approximate interpolation problem for $M(G)^{\hat{}}$. Theorem 2.3 shows that Sidon sets can be characterized in a manner similar to the characterization of $M(G)^{\hat{}-}$. Compare Theorem 2.3 to Theorem 5.4. Theorem 2.5 gives the topological characterization of Sidon sets.

3. An Example

3.1: In Section 5 we will show for E a Sidon set that $\chi_E \in M(G)^{\hat{}-}$. This implies by Corollary 2.6 that a finite union of Sidon sets is a Sidon set. In this section we show that there is a thin set of integers E with $\chi_E \in M(T)^{\hat{}-}$ where E is not a Sidon set.

3.2 Definition: For $f \in WAP(Z)$ (see 4.1.2), the **von Neumann mean,** $\mathscr{M}(f)$, is defined by

$$\mathscr{M}(f) = \lim_{N \to \infty} \frac{1}{2N+1} \sum_{n=-N}^{N} f(n).$$

3.3 Theorem: *There exists $F \subset Z_+$ (the set of nonnegative integers) such that $\chi_{F \cup -F} \in M(\mathbf{T})\hat{}^{-}$, $\mathscr{M}(\chi_{F \cup -F}) = 0$, but $\chi_F \notin WAP(Z)$ (thus $\chi_F \notin M(\mathbf{T})\hat{}^{-}$).*

Proof: Let $E = \{n_k\}_{k=1}^{\infty} \subset Z_+$ be such that $n_{k+1}/n_k \geq 3$. Let

$$F = \{n_i \pm n_j : i > j\}.$$

Consider the Riesz product

$$\prod_{k=1}^{\infty} (1 + \cos(n_k x)) = 1 + \sum_{k=1}^{\infty} c_k \cos(kx).$$

Now $c_k = 0$ unless

$$k \in S = \{n_{i_0} \pm n_{i_1} \pm \cdots \pm n_{i_m} : i_0 > i_1 > \cdots > i_m\}.$$

For $k = n_{i_0} \pm \cdots \pm n_{i_m}$ with $i_0 > \cdots > i_m$, $c_k = 1/2^m$. Thus $c_k = 1$ when $k \in E$, $c_k = \frac{1}{2}$ when $k \in F$, and $0 \leq c_k \leq \frac{1}{4}$ otherwise. It follows that there is $\mu \in M(\mathbf{T})$ such that $\hat{\mu}(n) = 1$ for $n \in E \cup -E \cup \{0\}$, $\hat{\mu}(n) = \frac{1}{2}$ for $n \in F \cup -F$ and $|\hat{\mu}(n)| \leq \frac{1}{4}$ otherwise. In particular, $\chi_{E \cup -E \cup \{0\}} \in M(\mathbf{T})\hat{}^{-}$. Thus

$$\chi_{F \cup -F} \in M(\mathbf{T})\hat{}^{-}.$$

Direct computation yields that $\mathscr{M}(\chi_{F \cup -F}) = 0$.

Let $(\chi_F)_n$ be defined by $(\chi_F)_n(m) = \chi_F(m + n)$. Now $(\chi_F)_{n_k} \xrightarrow{k} \chi_{E \cup -E}$ pointwise on Z. If $\chi_F \in WAP(Z)$, then there would exist a subsequence $\{n_i\} \subset \{n_k\}$ such that $(\chi_F)_{n_i} \xrightarrow{i} \chi_{E \cup -E}$ weakly, and hence quasi-uniformly (see 4.3.2; also [DS, p. 281]). But $(\chi_F)_{n_i} \not\xrightarrow{i} \chi_{E \cup -E}$ quasi-uniformly. This contradiction establishes the theorem. □

3.4 Corollary: *There exists $H \subset Z$ such that $\chi_H \in M(\mathbf{T})\hat{}^{-}$, $\mathscr{M}(\chi_H) = 0$, but H is not a Sidon set.*

Proof: Let $H = F \cup -F$ from Theorem 3.3. Suppose H is a Sidon set. Thus there is a $\mu \in M(\mathbf{T})$ such that $\hat{\mu} = 1$ on F and 0 on $-F$. Thus $\hat{\mu}\chi_H = \chi_F$ is in $M(\mathbf{T})\hat{}^{-}$. This is the required contradiction. Hence H is not a Sidon set. □

4. Sufficient Conditions

4.1 Remark: Let $E \subset \Gamma$. Now E is a Sidon set if and only if every countable subset of E is also a Sidon set. Hence we may restrict our attention to **countable** sets $E \subset \Gamma$, and therefore to countable groups Γ. In particular, in this case G will be metric, $C(G)$ separable, and thus the unit ball of $M(G)$ is sequentially weak-$*$ compact.

We now study sufficient conditions for Γ countable that will imply that a subset $E \subset \Gamma$ will be a Sidon set.

4.2 Definition: Let $E \subset \Gamma$ and $\gamma_1, \gamma_2, \ldots$ be an enumeration of the elements of E. Let $R_s(E, \gamma)$ denote the number of representations of γ in the form

$$\gamma = \pm \gamma_{n_1} \pm \gamma_{n_2} \pm \cdots \pm \gamma_{n_s}, n_1 < n_2 < \cdots < n_s.$$

4.3 Lemma: *Let $E \subset \Gamma$ and $1 \le B < \infty$ be such that $R_s(E,0) \le B^s, s = 1,2,\ldots$. Assume that $\gamma \in E$ and $2\gamma \ne 0$ implies $-\gamma \notin E$. Then*

$$\sum_{s=1}^{\infty} (2B)^{-s} R_s(E, \gamma) \le 2$$

for all $\gamma \in \Gamma$. In particular, $R_s(E, \gamma) \le 2(2B)^s$, $s = 1, 2, \ldots, \gamma \in \Gamma$.

Proof: Let $\beta = 1/2B$ and $\gamma_1, \gamma_2, \ldots$ be the elements of E. Let

$$f_k(x) = \begin{cases} 1 + \beta\gamma_k(x) + \beta\overline{\gamma_k(x)} & \text{if } 2\gamma_k \ne 0 \\ 1 + \beta\gamma_k(x) & \text{if } 2\gamma_k = 0, \end{cases}$$

and form the Riesz product

$$P_N(x) = \prod_{k=1}^{N} f_k(x).$$

Since $\beta \le \frac{1}{2}$ and $|\gamma_k(x)| = 1$, $P_N(x) \ge 0$. Now

$$P_N(x) = 1 + \sum_{\gamma\in\Gamma} c_N(\gamma)\gamma(x)$$

where

$$|c_N(\gamma)| \le \sum_{s=1}^{N} \beta^s \Sigma 1$$

where the inner summation runs over all $\gamma_{n_1}, \ldots, \gamma_{n_s}$ such that

$$\gamma = \pm \gamma_{n_1} \pm \cdots \pm \gamma_{n_s}, n_1 < \cdots < n_s.$$

Thus

$$|c_N(0)| \le \sum_{s=1}^{N} \beta^s R_s(E, 0) \le \sum_{s=1}^{N} (\beta B)^s \le 1.$$

Since $P_N \ge 0$, $\| P_N \|_1 = 1 + c_N(0) \le 2$. Thus $|\hat{P}_N(\gamma)| \le 2$ for all $\gamma \in \Gamma$. For $\gamma \ne 0$, $\hat{P}_N(-\gamma) = c_N(\gamma)$. Now fix $\gamma \in \Gamma$ and let $N \to \infty$. Thus

$$\lim_{N\to\infty} c_N(\gamma) = \sum_{s=1}^{\infty} \beta^s R_s(E, \gamma).$$

It follows that

$$\sum_{s=1}^{\infty} \beta^s R_s(E, \gamma) \leq 2$$

for all $\gamma \in \Gamma$. □

4.4 Theorem: *Suppose Γ is countable. Let $E \subset \Gamma$ and $0 < B < \infty$ be such that*

$$R_s(E, 0) \leq B^s \qquad (s = 1, 2, \ldots).$$

Then $E \cup (-E)$ is a Sidon set and $\chi_{E \cup (-E)} \in M(G)\hat{}^{-}$. In particular, $\chi_E \in M(G)\hat{}^{-}$.

Proof: We may assume $B \geq 1$ and $0 \notin E$. It follows from Corollary 2.6 that we may further assume that for $\gamma \in E$ with $2\gamma \neq 0$ that $-\gamma \notin E$.

By Lemma 4.3, there is a B such that $R_s(E, \gamma) \leq B^s$, $\gamma \in \Gamma$, $s = 1, 2, \ldots$. Let ϕ be a function on $E \cup (-E)$ such that $\phi(\gamma) = \pm 1$. Write $E = E_1 \cup E_2$ where

$$E_1 = \{\gamma : \gamma \in E \text{ and } \phi(\gamma) = \phi(-\gamma)\}$$

and

$$E_2 = \{\gamma : \gamma \in E \text{ and } \phi(\gamma) = -\phi(-\gamma)\}.$$

Let $\varepsilon > 0$ and $\beta = 1/(KB^2)$ where $K \geq 2$ and $2/(K - 1) < \varepsilon$. Define

$$g(\gamma) = \begin{cases} \beta\phi(\gamma) & \text{if } \gamma \in E_1, \\ i\beta\phi(\gamma) & \text{if } \gamma \in E_2. \end{cases}$$

Let $\gamma_1, \gamma_2, \ldots$ be the elements of E_1 and put

$$f_k(x) = \begin{cases} 1 + g(\gamma_k)\gamma_k(x) + \overline{g(\gamma_k)}\,(-\gamma_k)\,(x), & \text{if } 2\gamma_k \neq 0 \\ 1 + g(\gamma_k)\gamma_k(x), & \text{if } 2\gamma_k = 0. \end{cases}$$

Form the Riesz products

$$P_N(x) = \prod_{k=1}^{N} f_k(x). \qquad \text{Since } \beta = \tfrac{1}{2},\ P_N(x) \geq 0.$$

There is a subsequence of $\{P_N\}$ which converges weak-$*$ to a positive measure $\mu_1 \in M(G)$ such that

(a) $\| \mu_1 \| \leq \sup | \hat{P}_N(0) | \leq 1 + \sum_{2}^{\infty} \beta^s R_s(E, 0),$

(b) $| \hat{\mu}_1(\gamma_k) - g(\gamma_k) | \leq \sup_N | \hat{P}_N(\gamma_k) - g(\gamma_k) |$

$$\leq \sum_{2}^{\infty} \beta^s R_s(E, \gamma_k) \qquad \text{if } \gamma_k \in E_1,$$

(c) $|\hat{\mu}_1(-\gamma_k) - g(\gamma_k)| \le \sum_2^{\infty} \beta^s R_s(E, \gamma_k) \qquad \text{if } \gamma_k \in E_1,$

(d) $|\hat{\mu}_1(\gamma)| \le \sum_2^{\infty} \beta^s R_s(E, \gamma) \qquad \text{if } \gamma \notin E_1 \cup (-E_1) \cup \{0\},$

[e.g. R, p. 125].

Similarly for E_2 with μ_2. Now

$$\sum_2^{\infty} \beta^s R_s(E, \gamma) \le \sum_2^{\infty} (\beta B)^s = \frac{(\beta B)^2}{1 - \beta B} < \frac{1}{K(K-1)B^2}.$$

Let $\mu = \mu_1 - i\mu_2$. Then

$$|\hat{\mu}(\gamma) - \beta\phi(\gamma)| \le \frac{2}{B^2 K(K-1)} \qquad \text{if } \gamma \in E \cup (-E)$$

and

$$|\hat{\mu}(\gamma)| \le \frac{2}{B^2 K(K-1)} \qquad \text{if } \gamma \notin E \cup (-E) \cup \{0\}.$$

Let $\nu = \beta^{-1}\mu$. Then

$$|\hat{\nu}(\gamma) - \phi(\gamma)| \le \frac{2}{K-1} < \varepsilon \qquad \text{if } \gamma \in E \cup (-E),$$

and

$$|\hat{\nu}(\gamma)| \le \frac{2}{K-1} < \varepsilon \qquad \text{if } \gamma \notin E \cup (-E) \cup \{0\}.$$

Let $\lambda = \nu - \hat{\nu}(0)m$, where m is Haar measure. That E is a Sidon set now follows from Theorem 1.2. That $\chi_{E \cup (-E)} \in \overline{M(G)\hat{}}$ follows by letting $\phi = 1$. □

4.5 Remark: If there is $\gamma^* \in E$ such that $R_s(E, \gamma^*) \le B^s$, $s = 1, 2, \ldots$, then E satisfies the hypothesis of Theorem 4.4. For suppose

$$0 = \sum_1^{s} \pm \gamma_{n_k} \qquad \text{where } \gamma_{n_k} \in E,\ n_1 < n_2 < \cdots < n_s.$$

If $\pm\gamma^*$ appears in the above sum, then

$$\pm\gamma^* = \sum_1^{s-1} \pm\gamma_{n_k},\ n_1 < \cdots < n_{s-1}.$$

In this case,

$$R_s(E, 0) \le 2R_{s-1}(E, \gamma^*) \le 2B^{s-1}.$$

If $\pm y^*$ does not appear in the sum, then by adding γ^* to each side we have

$$\gamma^* = \sum_1^{s+1} \pm\gamma_{n_k},\ n_1 < n_2 < \cdots < n_{s+1}.$$

In this case, $R_s(E, 0) \leq R_{s+1}(E, \gamma^*) \leq B^{s+1}$. In either case, we would have $R_s(E, 0) \leq 2B^{s-1} + B^{s+1} \leq 3B^{s+1}$. $\square$

4.6 Corollary: *Let Γ be a countable discrete abelian group. Then E contains an infinite Sidon set.*

Proof: Choose an infinite independent subset, E of Γ. Then E is a Sidon set. $\square$

4.7 Corollary: *There exists $\delta > 0$ such that for any sequence of complex numbers, $\{c_n\}$, and for any k,*

$$\frac{\max | \pm c_1 \pm c_2 \pm \cdots \pm c_k |}{|c_1| + |c_2| + \cdots + |c_k|} \geq \delta > 0.$$

Proof: Let $G = \prod_{n=1}^{\infty} \{-1, 1\}_n$ and $\Gamma = \sum_{n=1}^{\infty} \{0, 1\}_n$. Let $E = \{y_n\}_{n=1}^{\infty}$ be a countable independent subset of Γ. Thus E is a Sidon set and there is a $\delta > 0$ such that for $0 \neq \lambda \in M(E)$,

$$\frac{\| \hat{\lambda} \|_\infty}{\| \lambda \|} \geq \delta.$$

Let λ_k be the measure on E given by $\lambda_k = c_1\delta_{y_1} + \cdots + c_k\delta_{y_k}$, where δ_{y_j} denotes point mass at y_j, $1 \leq j \leq k$. Now

$$0 < \delta \leq \frac{\| \hat{\lambda}_k \|_\infty}{\| \lambda_k \|} = \frac{\max | \pm c_1 \pm c_2 \pm \cdots \pm c_k |}{|c_1| + |c_2| + \cdots + |c_k|}. \quad \square$$

4.8 Definition: Let $E = \{n_k\}_{k=1}^{\infty} \subset Z_+$. If there is $q > 1$ such that $n_{k+1}/n_k \geq q$ for all $k = 1, 2, \ldots$, then E is called a **lacunary set.** The upper bound of the numbers λ which satisfy this inequality is called the degree of lacunarity.

4.9 Theorem: *Let $E = \{n_k\}_{k=1}^{\infty} \subset Z_+$ be a lacunary set with degree of lacunarity $q \geq 3$. Then E is a Sidon set and $\chi_E \in M(\mathbf{T})\hat{}^{-}$.*

Proof: Let $0 = n_{k_m} \pm n_{k_{m-1}} \pm \cdots \pm n_{k_1}$ where $k_m > k_{m-1} > \cdots > k_1$. Now

$$\begin{aligned} n_{k_m} \pm n_{k_{m-1}} \pm \cdots \pm n_{k_1} &\geq n_{k_m} - n_{k_{m-1}} - \cdots - n_2 - n_1 \\ &\geq n_{k_m}(1 - q^{-1} - q^{-2} - \cdots) \\ &\geq \frac{n_k q(q-2)}{q-1} \geq \frac{n_k q}{q-1} > 0 \end{aligned}$$

since $q \geq 3$. It follows that $R_s(E, 0) \leq 1$. The result now follows from Theorem 4.4. $\square$

4.10 Corollary: *If $E = \{n_k\}_{k=1}^{\infty} \subset Z_+$ is a lacunary set with degree of lacunarity $q > 1$, then E can be divided into a finite number of lacunary sequences in each of which the degree of lacunarity is not less than 3. In particular, if E is a finite union of lacunary sets, then E is a Sidon set and $\chi_E \in M(\mathbf{T})\hat{}^{-}$.*

Proof: Pick r such that $q^r \geq 3$. Divide the sequence $\{n_k\}$ into r sequences L_i $(i = 1, 2, \ldots, r)$ as follows:

$$L_i = \{n_i, n_{i+r}, \ldots, n_{i+jr}, \ldots\}.$$

Each term of $\{n_k\}$ belongs to one and only one of the sequences L_i, and for any i we have

$$\frac{n_{i+(j+1)r}}{n_{i+jr}} \geq q^r \geq 3 \; (j = 1, 2, \ldots).$$

Now use Theorem 4.4. □

4.11 Definition: Let $E = \{n_k\}_{k=1}^{\infty}$ be an increasing sequence of natural numbers, $n_1 < n_2 < \cdots < n_k < \cdots$, such that

$$\sum_{k=m}^{\infty} \frac{1}{n_k} = O\left(\frac{1}{n_m}\right).$$

We say that E is an L-set.

4.12 Corollary: *Let $E \subset Z_+$ be an L-set. Then E is a Sidon set and $\chi_E \in M(\mathbf{T})\hat{}^{-}$.*

Proof: It suffices to show that every L-set is a finite union of lacunary sets.

Let

$$r_m = \sum_{k=m}^{\infty} \frac{1}{n_k}.$$

Now there is C such that $r_m < C/n_m$ $(m = 1, 2, \ldots)$. Choose the number l such that l is the greatest integer less than $3C$. Now since the numbers $1/n_k$ decrease monotonically

$$(l+1)\frac{1}{n_{m+l}} \leq \sum_{k=m}^{m+l} \frac{1}{n_k} \leq \sum_{k=m}^{\infty} \frac{1}{n_k} \leq C\frac{1}{n_m}.$$

Thus

$$\frac{1}{n_{m+l}} \leq \frac{C}{l+1}\frac{1}{n_m} \leq \frac{1}{3}\frac{1}{n_m}; \qquad \text{that is, } \frac{n_{m+l}}{n_m} \geq 3.$$

Therefore the l series

$$L_1 = \{n_1, n_{1+l}, n_{1+2l}, \ldots\},$$
$$L_2 = \{n_2, n_{2+l}, n_{2+2l}, \ldots\}, \ldots,$$
$$L_l = \{n_l, n_{l+l}, n_{l+2l}, \ldots\}$$

form a partition of E and each is a lacunary set with degree of lacunarity $q \geq 3$. □

4.13 Remark: Not all Sidon subsets of Z can be expressed as finite unions of lacunary sets. For $m = 0, 1, 2, \ldots$, set $M = 2^m$, and let

$$E = \{3^{4M} + 3^{M+j}: j = 0, 1, \ldots, M - 1; m = 0, 1, 2, \ldots\}.$$

Now E contains M terms between $a_m = 3^{4M} + 3^M$ and $b_m = 3^{4M} + 3^{2M-1}$. Now $b_m < 2a_m$. Thus E is not a finite union of lacunary sets since any such union has a bounded number of elements between n and $2n$ as $n \to \infty$ [R, p. 127]. Finally, we show that $R_s(0, E) \leq 1$.

Suppose there is a representation of the form

$$0 = \pm n_1 \pm n_2 \pm \cdots \pm n_s \, (n_1 < n_2 < \cdots < n_s, n_i \in E, n_1 = 3^{4M} + 3^\alpha).$$

Now $n_i \equiv 0 \pmod{3^{\alpha+1}}$ for $2 \leq i \leq s$ and $n_1 \equiv 3^\alpha \pmod{3^{\alpha+1}}$. Thus $\pm 3^\alpha \equiv 0 \pmod{3^{\alpha+1}}$ and hence $\pm 1 \equiv 0 \pmod 3$ which is a contradiction.

4.14 Remark: A Sidon set cannot contain arbitrarily long arithmetic progressions. For suppose P is a Sidon set with the property that to each $s \in Z_+$, there is $p \in P$ and $q \in \Gamma$ with order of $q \geq s$ such that

$$P_s = \{p + q, p + 2q, \ldots, p + s\,q\} \subset P.$$

We may assume for $s_1 \neq s_2$ that $P_{s_1} \cap P_{s_2} = \emptyset$. Consider the measures

$$\lambda_s = \frac{1}{s} \sum_{n=1}^{s} e^{in \log(n)} \delta_{p+nq},$$

where δ_{p+nq} is the unit point measure at $p + nq$. Then $\| \lambda_s \| \leq 1$ and by a well-known but nontrivial inequality [Z, p. 199], we have for some $M < \infty$ that

$$\| \hat{\lambda}_s \|_\infty \leq \sup\left\{ \left| \frac{1}{s} \sum_{n=1}^{s} e^{in \log(n)} e^{i(p+nq)\theta} \right| : 0 \leq \theta \leq 2\pi \right\}$$
$$= \sup\left\{ \left| \frac{1}{s} \sum_{n=1}^{s} e^{in \log(n)} e^{inq\theta} \right| : 0 \leq \theta \leq 2\pi \right\}$$
$$\leq \frac{1}{s} M(s)^{\frac{1}{2}} = \frac{M}{(s)^{\frac{1}{2}}} \to 0 \text{ as } s \to \infty.$$

It follows from the definition of a Sidon set, that P cannot be a Sidon set. □

5. Uniformly Approximable Sidon Sets

5.1 Definition: Let E be a subset of Γ. If there is a function $f \in M(G)\hat{}^{-}$ such that $f = 1$ on E and $|f| \leq c < 1$ off E, then we say that E is a **peak set for** $M(G)\hat{}^{-}$. If such an f can be found in $M(G)\hat{}$, we say that E is a **peak set for $\mathbf{M(G)\hat{}}$.**

5.2 Lemma: *Let $E \subset \Gamma$ be a Sidon set. If E is a peak set for $M(G)\hat{}^{-}$, then E is a peak set for $M(G)\hat{}$.*

Proof: Since E is a Sidon set, there is a constant $C \geq 1$ such that if $\phi \in L^{\infty}(E)$, then there is $\mu \in M(G)$ such that $\hat{\mu} = \phi$ on E and $\|\mu\| \leq C\|\phi\|_{\infty}$.

That E is a peak set for $M(G)\hat{}^{-}$ is equivalent to the characteristic function of E, χ_E, belonging to $M(G)\hat{}^{-}$. Let $\nu \in M(G)$ be such that

$$\|\hat{\nu} - \chi_E\|_{\infty} < 1/4C.$$

Let $\mu \in M(G)$ be such that $\hat{\mu} = 1/\hat{\nu}$ on E and $\|\hat{\mu}\|_{\infty} \leq \|\mu\| \leq 2C$. Let $\lambda = \mu * \nu$. Then $\hat{\lambda} = 1$ on E and $|\hat{\lambda}(\gamma)| \leq 2C\,(1/4C) = \frac{1}{2}$ off E. □

5.3 Definition: Let $E \subset \Gamma$ be a Sidon set. We say that E is a **uniformly approximable Sidon set** if $\chi_E \in M(G)\hat{}^{-}$.

5.4 Theorem: *Let $E \subset \Gamma$. The following are equivalent:*

(A) *E is a uniformly approximable Sidon set.*

(B) *For $\varepsilon > 0$ and $f \in L^{\infty}(E)$, there is $\mu \in M(G)$ such that $\hat{\mu} = f$ on E and $|\hat{\mu}| < \varepsilon$ off E.*

(C) *If $f \in L^{\infty}(E)$, $\{\lambda_n\} \subset M(\Gamma)$ such that $\|\lambda_n\| \leq 1$ and $\hat{\lambda}_n \xrightarrow{n} 0$ pointwise on G, then $\Sigma_{\gamma \in \Gamma} f\, d\lambda_n \xrightarrow{n} 0$.*

Proof: That (A) and (B) are equivalent follows from Lemma 5.2.

That (A) and (C) are equivalent follows from the characterization of $M(G)\hat{}^{-}$ (see 3.3.12). □

5.5 Remarks: If E is a Sidon set and $\chi_E \in M(G)\hat{}$, then $C^B(E)$ forms a subalgebra of $M(G)\hat{}$. Now the set of Fourier–Stieltjes transforms contains no infinite dimensional B^*-subalgebra (see 3.2.1). Thus E must be finite.

If E is any subset of Z_+ with $\chi_E \in M(\mathbf{T})\hat{}$, then the theorem of F. and M. Riesz [R, p. 198] implies that χ_E is a transform of an absolutely continuous measure. It follows from the Riemann–Lebesgue lemma that E must be finite.

5.6 Theorem: *Let $E \subset \Gamma$ be such that $E = -E$ and E is a Sidon set. Then E is a uniformly approximable Sidon set.*

Proof: We may assume that $0 \notin E$. By the weak-$*$ compactness of the norm bounded sets in $M(G)$ it suffices to show for F a finite subset of E with $F = -F$ that there is

$$u \in L^1(G) \text{ with } \hat{u} = 1 \text{ on } F, |\hat{u}| \le 1/2 \text{ off } F \cup \{0\},$$

and $\| u \|_1 \le 8B^4$ where B is the Sidon-constant for E.

Let Ω be the finite group $\prod_{k=1}^{K} \{-1, 1\}_k$ written multiplicative where $F = \{\gamma_1, \ldots, \gamma_K\}$. For $\omega \in \Omega$, there is $\mu_\omega \in M(G)$, $\| \mu_\omega \| \le B$, $\hat{\mu}_\omega(\gamma_k) = \omega_k$, $1 \le k \le K$, where $\omega = (\omega_1, \ldots, \omega_K)$. For each $\omega \in \Omega$ put $g_\gamma(\omega) = \hat{\mu}_\omega(\gamma)$ and $f_\gamma = g_\gamma * g_\gamma$ (convolution on Ω). Thus for $\xi \in \Omega, f_\gamma(\xi) = \int_\Omega g_\gamma(\xi\lambda^{-1}) g_\gamma(\lambda)\, dm_\Omega(\lambda)$. Define $\nu_\omega \in M(G)$ by

$$\nu_\omega = \frac{1}{2^K} \sum_{\lambda \in \Omega} \mu_{\omega\lambda^{-1}} * \mu_\lambda.$$

Then $\| \nu_\omega \| \le B^2$, $\hat{\nu}_\omega(\gamma) = f_\gamma(\omega)$, $\hat{\nu}_\omega(\gamma_k) = \omega_k$, $1 \le k \le K$, and

$$\| \hat{f} \|_1 \le \| g_\gamma \|_2^2 \le \| g_\gamma \|_\infty^2 \le B^2, \gamma \in \Gamma.$$

Now define u_ω a continuous positive function on G by

$$u_\omega(g) = 8B^2 \prod_{k=1}^{K} \left(1 + \frac{\omega_k}{4B^2} R(\gamma_k(g)) \right), g \in G,$$

where

$R(\gamma_k(g)) = \frac{1}{2}(\gamma_k(g) + \overline{\gamma_k(g)})$ if $2\gamma_k \neq 0$ and $R(\gamma_k(g)) = \frac{1}{2}\gamma_k(g)$ if $2\gamma_k = 0$.

Let $u \in L^1(G)$ be defined by

$$u = \frac{1}{2^K} \sum_{\omega \in \Omega} \nu_\omega * u_\omega = \int_\Omega \nu_\omega * u_\omega \, dm_\Omega.$$

Now

$$\| u \|_1 \le \int_G \left(\int_\Omega \| \nu_\omega \| \| u_\omega \|_1 \, dm_\Omega(\omega)\right) dm_G \le B^2 \sup_{g \in G} \int_\Omega u_\omega(g)\, dm_\Omega(\omega) \le 8B^4$$

and

$$\hat{u}(\gamma_k) = \frac{1}{2^K} \sum_{\omega \in \Omega} \hat{\nu}_\omega(\gamma_k) \hat{u}_\omega(\gamma_k) = \int_\Omega \omega_k \hat{u}_\omega(\gamma_k)\, dm_\Omega(\omega) = 1, 1 \le k \le K.$$

Let $\gamma \in \hat{G}$, then

$$\begin{aligned} |\hat{u}(\gamma)| &= \left| \int_\Omega \hat{\nu}_\omega(\gamma) \hat{u}_\omega(\gamma)\, dm_\Omega(\omega) \right| = \left| \int_\Omega f_\gamma(\omega) \hat{u}_\omega(\gamma)\, dm_\Omega(\omega) \right| \\ &\le \| \hat{f}_\gamma \|_1 \sup_{\chi \in \hat{\Omega}} \left| \int \hat{u}_\omega(\gamma) \chi(\omega)\, dm_\Omega(\omega) \right| \\ &\le B^2 \sup_{\chi \in \hat{\Omega}} \left| \int_\Omega \hat{u}_\omega(\gamma) \chi(\omega)\, dm_\Omega(\omega) \right|. \end{aligned}$$

Let $\gamma \notin F \cup \{0\}$ and $\chi \in \hat{\Omega}$. It now suffices to show

$$\left| \int_\Omega \hat{u}_\omega(\gamma)\chi(\omega)\, dm_\Omega(\omega) \right| \le \frac{1}{2B^2}.$$

Write $\chi = (\varepsilon_1, \ldots, \varepsilon_K)$, $\varepsilon_k = 1$ or 0, $1 \le k \le K$. Then

$$\int_\Omega \hat{u}_\omega(\gamma)\chi(\omega)\, dm_\Omega(\omega) = \int_\Omega \hat{u}_\omega(\gamma) \prod_{k=1}^{K} \omega_k^{\varepsilon_k}\, dm_\Omega(\omega)$$

$$= 8B^2 \int_G \gamma(g)\, dm_G(g) \int_\Omega \prod_{l=1}^{K} \left(1 + \frac{\omega_l}{4B^2} R(\gamma_l(g))\right) \prod_{k=1}^{K} \omega_k^{\varepsilon_k}\, dm_\Omega(\omega)$$

$$= 8B^2 \int_G \gamma(g) \prod_{\varepsilon_l = 1} \frac{R(\gamma_l(g))}{4B^2}\, dm_G(g).$$

If $S = \{k : \varepsilon_k = 1\}$ has two or more elements,

$$\left| \int_\Omega \hat{u}_\omega(\gamma)\chi(\omega)\, dm_\Omega(\omega) \right| \le \frac{1}{2B^2}.$$

If $S = \emptyset$, then the integral is zero since $\gamma \ne 0$. If S has only one element, then the integral is zero since $\gamma \notin F \cup -F = F$. (Observe that this is the only time we have used the condition $F = -F$ and that without this condition we would deduce $|\hat{u}| \le \frac{1}{2}$ off $F \cup -F \cup \{0\}$.) □

5.7 Theorem: *Let $E \subset \Gamma$ be a Sidon set. Then E is a uniformly approximable Sidon set. Thus the union of two Sidon sets is a Sidon set.*

Proof: The second part of the theorem follows from the first part and Corollary 2.6.

Let $E \subset \Gamma$ be a Sidon set. Let $H = \mathbf{T} \oplus G$ and thus $\hat{H} = Z \oplus \Gamma$. Let $E_0 = \{(0, \gamma) \in \hat{H} : \gamma \in E\}$ and $E_1 = \{(1, \gamma) \in \hat{H} : \gamma \in E\}$. Since E is a Sidon set, E_0 is a Sidon set in $\hat{H}$, and thus E_1 is a Sidon set in $\hat{H}$.

By repeating the proof of Theorem 5.6 one shows that there is $\mu \in M(H)$ with $\hat{\mu} = 1$ on E_1, $|\hat{\mu}| \le \frac{1}{2}$ off $E_1 \cup -E_1 \cup \{0\}$ and $\|\mu\| \le 8B^4$. Let $\nu \in M(H)$ with $\hat{\nu} = 1$ on $\Gamma_1 = \{(1, \gamma) : \gamma \in \Gamma\}$ and $\hat{\nu} = 0$ off Γ_1. Now $\mu * \nu \in M(H)$ is such that $(\mu * \nu)\hat{} = 1$ on E_1, $(\mu * \nu)\hat{} = 0$ off Γ_1, $|(\mu * \nu)\hat{}| \le \frac{1}{2}$ on $\Gamma_1 \backslash E_1$, and $\|\mu * \nu\| \le 8B^4$. Now let $\sigma \in M(H)$ be such that $\hat{\sigma}(\chi) = \hat{\mu}(\chi - (1, 0))$, $\chi \in \hat{H}$. Thus $\sigma \in M(H)$, $\hat{\sigma} = 1$ on E_0, $\hat{\sigma} = 0$ off $\Gamma_0 = \{(0, \gamma) : \gamma \in \Gamma\}$, $|\hat{\sigma}| \le \frac{1}{2}$ on $\Gamma_0 \backslash E_0$, and $\|\sigma\| \le 8B^4$. Finally, for $\hat{\lambda}$ a Fourier–Stieltjes transform supported on Γ_0, the annihilator of $\mathbf{T}_0 = \mathbf{T} \oplus \{0\} \subset H$, there is $\rho \in M(G) \cong M(H/\mathbf{T}_0)$ with $\|\rho\| \le \|\lambda\|$ and $\hat{\rho}(\gamma) = \hat{\lambda}((0, \gamma))$, $\gamma \in \Gamma$ (recall [R, p. 53]). □

6. Historical Notes

6.1: A proof of the fundamental characterization of Sidon sets [Theorem 1.3] can be found in [R, p. 121]. See also Banach [1, p. 218], Hewitt and Zuckerman [1, p. 310], Kahane [1, p. 310], and Rudin [4, p. 207]. Theorem 2.1 is due to Rudin [4, p. 208]. Theorem 2.3 is in Ramirez [1, p. 331]. Corollary 2.4 is from Ramirez [2, p. 616]. It is related to a result for lacunary series [BII, p. 249]. Theorem 3.3 and Corollary 3.4 are in Ramirez [2]. Theorem 3.3 yields information on the ideal structure of the weakly almost periodic functions on Z [BH, p. 148]. Theorem 4.4 and Remark 4.5 give the best sufficient condition we know for a set to be a Sidon set, Rider [1, p. 390]. Rider's theorem is a generalization of the work of Rudin [4, p. 209], Hewitt and Zuckerman [1, p. 5], and Stečkin [1, p. 394]. Sidon [1, p. 207] showed that a finite union of lacunary sets is a Sidon set (Corollary 4.10). The relationship between L-sets and lacunary sets is taken from [BI, p. 8]. The interesting example of a Sidon set which is not a finite union of lacunary sets (Remark 4.13) is due to Hewitt and Zuckerman [1, p. 8]. Remark 4.14 is taken from Ramirez [1, p. 332]. A stronger result can be found in Rudin [4, p. 214], Kahane [1, p. 312], and [K, p. 58]. For a survey article on Riesz products see Keogh [1].

For further results on lacunary Fourier series see [E, Chapter 15]. Peak sets were studied in Ramirez [2]. Uniformly approximable Sidon sets are characterized by Chaney [1]. Theorems 5.6 and 5.7 are due to Drury [1].

CHAPTER 6

IDEMPOTENT MEASURES

1. Introduction

1.1: Here we present the elegant new proof by T. Itô and I. Amemiya [1] of the idempotent measures theorem of Helson [1], Rudin [1], and P. J. Cohen [1]. Helson proved the theorem for the group $\mathbf{T}$, Rudin did it for $\mathbf{T}^n$, and Cohen did the general LCA case. A measure is said to be idempotent if $\mu * \mu = \mu$, and this implies that $\hat{\mu}$ takes only the values zero and one. The theorem states that $S(\mu) = \{\gamma \in \Gamma : \hat{\mu}(\gamma) = 1\}$ is in the coset-ring of Γ, which is the least ring of subsets (closed under finite unions and complementation) containing the cosets of open subgroups of Γ. We also give the norm inequality proved by S. Saeki [1] which states that if μ is idempotent and $\| \mu \| > 1$, then $\| \mu \| \geq (1 + \sqrt{2})/2$, a result suggested by Rudin.

1.2: If Λ is an open subgroup of Γ and H is its annihilator, that is, $H = \{x \in G : (x, \gamma) = 1, \text{ all } \gamma \in \Lambda\}$, then $\hat{H} \cong \Gamma/\Lambda$ discrete, so H is compact. For $\gamma_0 \in \Gamma$, let $\mu = \gamma_0 \cdot m_H$ (notation: m_H is the normalized Haar measure on H considered as a measure on G; if $f \in C(G)$, $\lambda \in M(G)$ then $f \cdot \lambda$ is the measure $f\, d\lambda$) then $S(\mu) = \Lambda + \gamma_0$, and μ is idempotent. The main theorem then says that any idempotent is a finite linear combination with integer coefficients of such $\gamma_0 \cdot m_H$.

1.3 A Useful Extension: Let $F(G) = \{\mu \in M(G) : \hat{\mu} \text{ is integer-valued}\}$. This contains the idempotents, is closed under addition, convolution, and multiplication by characters, and if $\mu_1, \mu_2 \in F(G)$, $\mu_1 \neq \mu_2$, then $\| \mu_1 - \mu_2 \| \geq 1$.

If $\mu \in F(G)$ then the range of $\hat{\mu}$ is finite since $\| \hat{\mu} \|_\infty \le \| \mu \|$. It will suffice to consider $F(G)$ for compact groups G by the following:

1.4 Theorem: *If $\mu \in F(G)$ then the support group of μ (the least closed subgroup whose complement is μ-null) is compact.*

Proof: Assume $\mu \ne 0$. Let H be the support group of μ, then μ may be regarded as an element of $F(H)$. Let γ_1, γ_2 be different elements of $\hat{H}$, then $\gamma_1 \cdot \mu \ne \gamma_2 \cdot \mu$ or else $|\mu| \{x \in H:(x, \gamma_1) \ne (x, \gamma_2)\} = 0$, so μ is concentrated on $\{x \in H:(x, \gamma_1 - \gamma_2) = 1\}$ a closed proper subgroup of H, a contradiction. Thus if $\gamma \ne 0$ then $\mu, \gamma \cdot \mu \in F(H)$, $\mu \ne \gamma \cdot \mu$, so $\| \mu - \gamma \cdot \mu \| \ge 1$. There exists a compact $K \subset H$ such that $|\mu| (H \backslash K) < 1/4$. Also $V = \{\gamma \in \hat{H}: |1 - (x, \gamma)| < 1/(3 \| \mu \|)$, all $x \in K\}$ is open in Γ, and for $\gamma \in V$

$$\| \mu - \gamma \cdot \mu \| \le \int_H |1 - (x, \gamma)| \, d|\mu|(x) \le \int_K + \int_{H \backslash K} \le 1/3 + 1/2 < 1.$$

Thus $V = \{0\}$, $\hat{H}$ is discrete and H is compact. □

2. The Cohen Idempotent Theorem

2.1: Henceforth we assume that G is **compact,** then $F(G)$ is closed in the weak-$*$ topology. For $\mu, \nu \in M(G)$, we write $\mu \perp \nu$ to mean μ and ν are mutually singular.

2.2 Helson's Translation Lemma: *Let $\mu \in M(G)$, G compact, $\mu = \mu_a + \mu_s$, $\mu_a \ll m_G$, $\mu_s \perp m_G$, and suppose ν is a weak-$*$ cluster point of $\{\gamma \cdot \mu : \gamma \in \Gamma\}$, then $|\nu E| \le |\mu_s| E$ for any Borel set $E \subset G$, in particular $\nu \perp m_G$.*

Proof: Let Ω be the collection of weak-$*$ neighborhoods of ν ordered by inclusion, and for each $\alpha \in \Omega$ let $\gamma_\alpha \in \Gamma$ such that $\gamma_\alpha \cdot \mu \in \alpha \backslash \{\nu\}$. Then $\{\gamma_\alpha \cdot \mu\}$ is a generalized sequence and we will show that $\gamma_\alpha \cdot \mu_a \xrightarrow{\alpha} 0$ (weak-$*$). Observe that for any finite set $S \subset \Gamma$ there exists $\alpha \in \Omega$ such that $\beta \subset \alpha$ implies $\gamma_\beta \notin S$. Now let $\varepsilon > 0$, then there exists finite $S \subset \Gamma$ such that $\gamma \notin S$ implies $|\hat{\mu}_a(\gamma)| < \varepsilon$ (by absolute continuity). Let $\gamma_0 \in \Gamma$, then there exists $\alpha \in \Omega$ such that $\beta \subset \alpha$ implies $\gamma_\beta \notin S - \gamma_0$. Then

$$\left| \int_G \gamma_0 \gamma_\beta \, d\mu_a \right| = |\hat{\mu}_a(\gamma_\beta + \gamma_0)| < \varepsilon,$$

hence $\lim_\alpha \int_G \gamma_0 \gamma_\alpha \, d\mu_a = 0$. This immediately extends to trigonometric polynomials, and hence to all of $C(G)$.

Thus $\gamma_\alpha \cdot \mu_s \xrightarrow{\alpha} \nu$ (weak-$*$), and for any open $V \subset G$, $f \in C(G)$ such that $\operatorname{spt} f \subset V$ and $\| f \|_\infty \le 1$, we have

$$\left| \int_V f \, d\nu \right| = \left| \lim_\alpha \int_V f \gamma_\alpha \, d\mu_s \right| \le |\mu_s| V$$

Thus $|\nu| V \le |\mu_s| V$ and regularity extends this to all Borel sets. □

2.3 Lemma: *If $\mu \in M(G)$, and ν is a weak-$*$ cluster point of $A = \{\gamma \cdot \mu : \gamma \in \Gamma\}$, and H is a compact subgroup of G, then $\nu|^H = (\gamma_0 \cdot \mu)|^H$ for some $\gamma_0 \cdot \mu \in A$ or $\nu \perp m_H$.*

Proof: There exists a generalized sequence $\{\gamma_\alpha \cdot \mu\} \subset A$ such that $\gamma_\alpha \cdot \mu \xrightarrow{\alpha} \nu$. Then $(\gamma_\alpha \cdot \mu)|^H \xrightarrow{\alpha} \nu|^H$, and if $\nu|^H \neq (\gamma_\alpha \cdot \mu)|^H$ for some α, then $\nu|^H$ is a weak-$*$ cluster point (in $M(H)$) and by Lemma 2.2 $\nu \perp m_H$ (for $\varepsilon > 0$, any $f \in C(H)$ extends to $g \in C(G)$ with $\int_{G \setminus H} |g| \, d(|\mu| + |\nu|) < \varepsilon$ by regularity). □

2.4 Lemma: *Let G be compact, $\mu \in F(G)$, $\mu \neq 0$, let ν be a weak-$*$ cluster point of $A \subset \{\gamma \cdot \mu : \gamma \in \Gamma\}$ then*

$$\|\nu\| \leq \|\mu\| - 1/(12\|\mu\|).$$

Proof: Assume $\nu \neq 0$, then $0 < \|\nu\| \leq \|\mu\|$ since closed bounded sets are weak-$*$ closed. Choose $K > 0$ such that $K < \|\nu\|/\|\mu\|$, then there exists $f \in C(G)$ with $\|f\|_\infty \leq 1$ and $\int_G f \, d\nu > K\|\mu\|$. Then $V = \{\lambda \in M(G): \mathrm{Re}(\int_G f \, d\lambda) > K\|\mu\|\}$ is weak-$*$ open and a neighborhood of ν, thus there exists $\gamma_1 \cdot \mu$, $\gamma_2 \cdot \mu \in A \cap V$, with $\gamma_1 \cdot \mu \neq \gamma_2 \cdot \mu$. Let $M_j = \mathrm{Re}(\int_G f\gamma_j \, d\mu)$, $j = 1, 2$, $d\mu = \theta \, d|\mu|$, θ measurable and $|\theta| = 1$, and $f\gamma_j\theta = g_j + ih_j$ (note $|f\gamma_j\theta| \leq 1$). Then

$$\begin{aligned}\|\mu\| &\geq \left|\int_G (g_j + i|h_j|) \, d|\mu|\right| \\ &= \{(\textstyle\int_G g_j \, d|\mu|)^2 + (\int_G |h_j| \, d|\mu|)^2\}^{1/2} \\ &= \{M_j^2 + (\textstyle\int_G |h_j| \, d|\mu|)^2\}^{1/2},\end{aligned}$$

thus

$$\int_G |h_j| \, d|\mu| \leq (\|\mu\|^2 - M_j^2)^{1/2}$$

(a method of Cohen [1, p. 206]).
Now

$$\begin{aligned}\int_G |1 - f\gamma_j\theta| \, d|\mu| &\leq \int_G (1 - g_j) \, d|\mu| + \int_G |h_j| \, d|\mu| \\ &\leq \|\mu\| - M_j + (\|\mu\|^2 - M_j^2)^{1/2} \\ &\leq \|\mu\| \{1 - K + (1 - K^2)^{1/2}\} \text{ for } j = 1, 2.\end{aligned}$$

Since $\gamma_1 \cdot \mu \neq \gamma_2 \cdot \mu$, we have

$$\begin{aligned}1 &\leq \|\gamma_1 \cdot \mu - \gamma_2 \cdot \mu\| \leq \int_G |\gamma_1 - \gamma_2| \, d|\mu| \\ &\leq \int_G (|\gamma_1 - \gamma_1\gamma_2 f\theta| + |\gamma_1\gamma_2 f\theta - \gamma_2|) \, d|\mu| \\ &\leq 2\|\mu\| \{1 - K + (1 - K^2)^{1/2}\}.\end{aligned}$$

Hence $1 - K + (1 - K^2)^{1/2} \geq 1/(2\|\mu\|)$. The left-hand side is a strictly decreasing function of K, which is 2 for $K = 0$, and 0 for $K = 1$; thus it is equivalent to an inequality of the form $K \leq K_0 < 1$, where K_0 is the solution of the quadratic

$$2K_0^2 - 2K_0\{1 - 1/(2\|\mu\|)\} + 1/(2\|\mu\|)^2 - 1/\|\mu\| = 0$$

(and of the inequality $K_0 \geq 1 - 1/(2\|\mu\|)$). By using the Newton method at the point $K = 1$ we obtain $K_0 \leq 1 - 1/(12\|\mu\|^2)$. Thus $K \leq 1 - 1/(12\|\mu\|^2)$ but this holds for all $K < \|\nu\|/\|\mu\|$ hence $\|\nu\| \leq \|\mu\| - 1/(12\|\mu\|)$. □

2.5 Definition: We say $\mu \in F(G)$ is **canonical** if and only if

$$\mu = \sum_{i=1}^{m} n_i \gamma_i \cdot m_H, \quad \text{for} \quad n_i \in Z, \quad \gamma_i \in \Gamma$$

and m_H the Haar measure of H, a compact subgroup of G.

2.6 Theorem: *Each $\mu \in F(G)$ is a finite sum of mutually singular canonical measures.*

Proof: We may assume G is compact. Let $\mu \in F(G)$, $\mu \neq 0$, and let

$$A = \{\gamma \cdot \mu : \gamma \in \Gamma, \quad \textstyle\int_G \gamma \, d\mu \neq 0\}.$$

Let $\overline{A}$ be the weak-$*$ closure of A, then $0 \notin \overline{A}$, since

$$\left| \textstyle\int_G 1 \, d(\gamma \cdot \mu) \right| \geq 1 \quad \text{for all} \quad \gamma \cdot \mu \in A.$$

Let $\delta = \inf\{\|\lambda\| : \lambda \in \overline{A}\}$, then for each $n = 1, 2, \ldots$ the set $\{\lambda \in \overline{A} : \|\lambda\| \leq \delta + 1/n\}$ is weak-$*$ compact, thus by the finite intersection property there exists $\nu \in \overline{A}$ with $\|\nu\| = \delta$, and so $\delta > 0$. Let $B = \{\gamma \cdot \nu : \gamma \in \Gamma, \int_G \gamma \, d\nu \neq 0\}$, then $B \subset \overline{A}$, for if $\gamma_\alpha \cdot \mu \xrightarrow{\alpha} \nu$ (weak-$*$), then $\gamma\gamma_\alpha \cdot \mu \xrightarrow{\alpha} \gamma \cdot \nu$, and since $\int_G \gamma \, d\nu \neq 0$, there exists a convergent subnet $\{\gamma_\beta \cdot \mu\}$ such that $\int_G \gamma\gamma_\beta \, d\mu \neq 0$, thus $\gamma \cdot \nu \in \overline{A}$. But now B is finite, for if σ were a weak-$*$ cluster point of B, then $\|\sigma\| < \|\nu\|$ by Lemma 2.4, and $\sigma \in \overline{A}$, contradicting the norm minimality of ν in $\overline{A}$.

We now construct the support group of ν. Define an equivalence relation on $S = \{\gamma \in \Gamma : \int_G \gamma \, d\nu \neq 0\}$ by $\gamma_1 \sim \gamma_2$ if and only if $\gamma_1 \cdot \nu = \gamma_2 \cdot \nu$. Let $\theta_1, \ldots, \theta_m$ be an enumeration of the equivalence classes, then for each i define

$$H_i = \{x \in G : (x, \gamma_1) = (x, \gamma_2) \text{ for all } \gamma_1, \gamma_2 \in \theta_i\}.$$

So H_i is a compact subgroup of G and $|\nu|(H_i)^c = 0$. Let $H = \bigcap_{i=1}^m H_i$, a compact subgroup, and $|\nu| H^c = 0$. For each i, define $\chi_i \in \hat{H}$ by $\chi_i = \gamma|_H$ for some $\gamma \in \theta_i$; then χ_i depends only on i, and for $i \neq j$, $\chi_i \cdot \nu \neq \chi_j \cdot \nu$, thus $\chi_i \neq \chi_j$. Now let $\chi \in \hat{H}$ with $\int_H \chi \, d\nu \neq 0$, then χ is the restriction to H of some $\gamma_0 \in S$, hence $\chi = \chi_i$ for some i. Thus ν has only a finite number of nonzero

Fourier coefficients on H, and is therefore the sum

$$\sum_{i=1}^{m} n_i\bar{\chi}_i \cdot m_H = \sum_{i=1}^{m} n_i\bar{\gamma}_i \cdot m_H, \qquad \text{for } n_i \in Z, \qquad \gamma_i \in \theta.$$

If ν is not a cluster point of A, then $\nu = \gamma_0 \cdot \mu$ for some $\gamma_0 \in \Gamma$, so $\mu = \bar{\gamma}_0 \cdot \nu$ canonical. Otherwise ν is a cluster point of A and by Lemma 2.3 since $\nu \ll m_H$, we have that $\nu = \gamma_0 \cdot \mu|^H$ for some $\gamma_0 \in \Gamma$. Let $\mu_1 = \bar{\gamma}_0 \cdot \nu$, and $\mu = \mu_1 + (\mu - \mu_1)$, then $\mu_1 \perp (\mu - \mu_1)$ (since $\mu_1 = \mu|_H$) and $\| \mu \| = \| \mu_1 \| + \|\mu - \mu_1 \| \geq 1 + \| \mu - \mu_1 \|$, so $\| \mu - \mu_1 \| \leq \| \mu \| - 1$. The same argument can now be applied to $\mu - \mu_1$, and since the norm is decreased by at least one, this process must stop after a finite number of steps. □

3. Norms of Idempotent Measures

3.1: Rudin showed that an idempotent measure μ has $\| \mu \| = 1$ if and only if $S(\mu)$ is an open coset in Γ, and that otherwise (if $\mu \neq 0$) $\| \mu \| > \sqrt{5}/2 \doteqdot 1.118$. Further he showed that if $S(\mu)$ is a union of two cosets of the same open subgroup then $\| \mu \| \geq (1 + \sqrt{2})/2 \doteqdot 1.207$. Saeki [1] has shown that in fact if μ is idempotent and $\| \mu \| > 1$ then $\| \mu \| \geq (1 + \sqrt{2})/2$.

Except in 3.2, G (as in Section 2) is a **compact** group.

3.2 Proposition: *Let μ be an idempotent measure on G, then $\| \mu \| = 1$ if and only if $S(\mu)$ is an open coset.*

Proof: If $S(\mu)$ is an open coset, then $\mu = \gamma \cdot m_H$ for some compact subgroup H of G, some $\gamma \in \Gamma$ and $\| \mu \| = 1$.

Now suppose $\| \mu \| = 1$, then for some $\gamma \in \Gamma$, $\hat{\mu}(\gamma) = 1$. Let $\sigma = \gamma \cdot \mu$, then $\| \sigma \| = 1$ and $\hat{\sigma}(0) = 1$, that is, $\| \sigma \| = \int_G d\sigma$, so $\sigma \geq 0$. Then $\hat{\sigma}$ is positive definite on Γ, so if $\gamma \in S(\sigma)$, then $\hat{\sigma}(-\gamma) = \hat{\sigma}(\gamma) = 1$ and $-\gamma \in S(\sigma)$. Recall the Krein inequality for positive definite functions:

$$|\hat{\sigma}(\gamma_1) - \hat{\sigma}(\gamma_2)|^2 \leq 2\,\hat{\sigma}(0)\,\mathrm{Re}\,\{\hat{\sigma}(0) - \hat{\sigma}(\gamma_1 - \gamma_2)\}.$$

Let $\gamma_1, \gamma_2 \in S(\sigma)$, then $-\gamma_2 \in S(\sigma)$ and

$$|\hat{\sigma}(\gamma_1 - \gamma_2) - \hat{\sigma}(\gamma_1)|^2 \leq 2\,\{\hat{\sigma}(0) - \hat{\sigma}(-\gamma_2)\} = 0,$$

thus $\gamma_1 - \gamma_2 \in S(\sigma)$, and $S(\sigma)$ is an open subgroup of Γ. Finally $S(\mu)$ is a coset of $S(\sigma)$. □

3.3 Lemma: *Let $f \in C(\mathbf{T})$, $\gamma \in \Gamma$, then $\int_G f((x, \gamma))\, dx$ equals $1/2\pi \int_0^{2\pi} f(e^{i\theta})\, d\theta$ if the order of γ is infinite, and equals $(1/p) \sum_{r=1}^{p} f(e^{2\pi ri/p})$ if the order of γ is p.*

Proof: If the order of γ is p, then the range of γ is $\{e^{2\pi ir/p}: r = 1, \ldots, p\}$ and

$$\int_G f((x, \gamma))\, dx = \sum_{r=1}^{p} f(e^{2\pi ir/p})\, m_G(\gamma^{-1}(e^{2\pi ir/p})).$$

Each $\gamma^{-1}(e^{2\pi ir/p})$ is a coset of the kernel of γ, and all have equal Haar measure, hence $1/p$.

If the order of γ is infinite, then the range of γ is all of $\mathbf{T}$ (since the range of γ is an infinite compact subgroup of $\mathbf{T}$). The linear functional on $C(\mathbf{T})$ defined by $f \mapsto \int_G f((x, \gamma))\, dx$ is easily seen to be translation-invariant, and is normalized, so by the uniqueness of Haar measure on $\mathbf{T}$, $\int_G f((x, \gamma))\, dx = 1/2\pi \int_0^{2\pi} f(e^{i\theta})\, d\theta$. □

3.4 Theorem: *Let* $\mu = (\gamma_1 + \gamma_2) \cdot m_G$, $\gamma_1 \neq \gamma_2$, *then* $\| \mu \| \geq (1 + \sqrt{2})/2$.

Proof: It suffices to compute $\| 1 + \gamma \|_1$ for $\gamma \in \Gamma$. If the order of γ is infinite, then

$$\begin{aligned} \| 1 + \gamma \|_1 &= \int_G | 1 + (x, \gamma) |\, dx = 1/2\pi \int_0^{2\pi} | 1 + e^{i\theta} |\, d\theta \\ &= 4/\pi. \end{aligned}$$

If the order of γ is p then

$$\begin{aligned} \| 1 + \gamma \|_1 &= 1/p \sum_{r=1}^{p} | 1 + e^{2\pi ir/p} | \\ &= 2/p \sum_{r=1}^{p} | \cos(\pi r/p) | \\ &= \begin{cases} 2/(p \tan \pi/2p), & p \text{ even} \\ 2/(p \sin \pi/2p), & p \text{ odd.} \end{cases} \end{aligned}$$

Now $2/(p \sin \pi/2p)$ decreases to $4/\pi$ as $p \to \infty$, and $2/(p \tan \pi/2p)$ increases to $4/\pi$. The case $p = 2$ gives norm one, since $(1 + \gamma) \cdot m_G$ is then the Haar measure of kernel (γ), thus the minimum norm greater than 1 is achieved at $p = 4$, where $\| 1 + \gamma \|_1 = (1 + \sqrt{2})/2$. □

3.5 Theorem: *If* μ *is an idempotent measure on a compact group* G *and* $\| \mu \| > 1$ *then* $\| \mu \| \geq (1 + \sqrt{2})/2$.

Proof: We may assume that $0 \in S(\mu)$ (or else consider $\gamma \cdot \mu$ for a suitable γ), and by Proposition 3.2, $S(\mu)$ is not a subgroup of Γ, so there exists $\gamma_0, \gamma_1 \in S(\mu)$ such that $\gamma_1 - \gamma_0 \notin S(\mu)$.

Suppose first that $2\gamma_0 \in S(\mu)$, and let

$$f(x) = (x, \gamma_0)[1 + \operatorname{Re}(x, \gamma_0)] + (x, \gamma_1)[1 - \operatorname{Re}(x, \gamma_0)].$$

Then

$$|f(x)| \leq |1 + \operatorname{Re}(x, \gamma_0)| + |1 - \operatorname{Re}(x, \gamma_0)| = 2$$

and

$$\begin{aligned}\int_G f\,d\mu &= \hat{\mu}(\gamma_0) + \tfrac{1}{2}[\hat{\mu}(2\gamma_0) + \hat{\mu}(0)]\\ &\quad + \hat{\mu}(\gamma_1) - \tfrac{1}{2}[\hat{\mu}(\gamma_1 + \gamma_0) + \hat{\mu}(\gamma_1 - \gamma_0)]\\ &= 3 - \tfrac{1}{2}\hat{\mu}(\gamma_1 + \gamma_0) \geq 5/2.\end{aligned}$$

Thus $\| f \|_\infty \| \mu \| \geq 5/2$, and $\| \mu \| \geq 5/4$.

Suppose now that $2\gamma_0 \notin S(\mu)$. Let Γ_0 be the cyclic group generated by γ_0, and m_0 the Haar measure of the annihilator of Γ_0. Then if $\mu_0 = \mu * m_0$, $\| \mu_0 \| \leq \| \mu \|$, and μ_0 is idempotent so it suffices to consider μ_0. Further $S(\mu_0) = S(\mu) \cap \Gamma_0$. Let $J = \{n \in Z\colon n\gamma_0 \in S(\mu_0)\}$, then observe that $0, 1 \in J$ but $2 \notin J$. At least one of the following conditions holds:

(a) $S(\mu_0)$ is a union of two cosets of a subgroup Λ of Γ_0.
(b) J contains three successive integers.
(c) J contains an isolated integer, that is, some $p \in J$, but $p - 1, p + 1 \notin J$.
(d) None of (a), (b), (c) holds.

CASE (a): Since $2\gamma_0 \notin S(\mu_0)$, the two cosets are Λ and $\Lambda + \gamma_0$, $\mu_0 = (1 + \gamma_0) \cdot m_1$ where m_1 is the Haar measure of the annihilator of Λ. Also $S(\mu_0)$ is not a subgroup so by Theorem 3.4 $\| \mu_0 \| \geq (1 + \sqrt{2})/2$. Observe that this condition is equivalent to J being of the form $\{kp\colon k \in Z\} \cup \{kp + 1\colon k \in Z\}$ for some $p \geq 3$.

CASE (b): There exists $q \in J$ such that $q + 1$, $q + 2 \in J$ but $q - 1$ or $q + 3 \notin J$. (Details: Let p, $p + 1$, $p + 2 \in J$; if $p > 2$ and $p - 1 \in J$ then $p > 3$, and we replace p by $p - 1$, by translating μ_0, and this process must stop since $2 \notin J$; if $p < 0$, and $p + 3 \in J$, then $p < -1$, and we replace p by $p + 1$, etc.).

If $q - 1 \notin J$, let

$$f(x) = (x, (q + 1)\gamma_0)[1 + \operatorname{Re}(x, \gamma_0)] + (x, q\gamma_0)[1 - \operatorname{Re}(x, \gamma_o)],$$

then $\| f \|_\infty \leq 2$, $\int_G f\,d\mu_0 = 5/2$, thus $\| \mu_0 \| \geq 5/4$.

If $q + 3 \notin J$, let

$$f(x) = (x, (q + 1)\gamma_0)[1 + \operatorname{Re}(x, \gamma_0)] + (x, (q + 2)\gamma_0)[1 - \operatorname{Re}(x, \gamma_0)]$$

and as above we show that $\| \mu_0 \| \geq 5/4$.

CASE (c): There exists $q \in J$, such that $q - 1$, $q + 1 \in J$. Let $f(x) = (1 + \operatorname{Re}(x, \gamma_0)) + (x, q\gamma_0)[1 - \operatorname{Re}(x, \gamma_0)]$. Then $\| f \|_\infty \leq 2$ and

$$\int_G f\,d\mu_0 = 5/2 + \hat{\mu}_0(-\gamma_0) \geq 5/2$$

so $\| \mu_0 \| \geq 5/4$.

CASE (d): Let $J_0 = \{p \in J\colon p + 1 \in J\}$; since (c) does not hold, $J = J_0 \cup (J_0 + 1)$; since (b) does not hold, $p, q \in J_0$ and $p \neq q$ implies $|p - q| \geq 3$;

and since (a) does not hold J_0 is not a group. For each $p \in J_0$ let p_+ be the next larger element of J_0, or ∞ if there is none, and let p_- be the next smaller element of J_0 or $-\infty$ if there is none. Then there exists $p \in J_0$ such that $q = p_+ - p \neq p - p_- = r$. Now if $r > q$, then q is finite, and we translate J by $-p$ to obtain $q, q + 1 \in J$ and $-s \notin J$ for $1 \leq s < q\,(\leq r - 1)$. If $r < q$, then r is finite, and we replace J by $-J$ (that is, $\mu_0(-x)$) and translate by $p + 1$ to obtain r, $r + 1 \in J$ and $-s \notin J$ for $1 \leq s < r\ (\leq q - 1)$. Thus we may assume that there exists $q \in J$, $q \geq 3$, such that q, $q + 1 \in J$ and $-s \notin J$ for $1 \leq s < q$. Then let

$$f(x) = [1 + \mathrm{Re}\,(x, (q + 1)\gamma_0)] + (x, \gamma_0)\,[1 - \mathrm{Re}\,(x, (q + 1)\gamma_0)].$$

Then $\| f \|_\infty \leq 2$, and

$$\textstyle\int_G f\, d\mu_0 = 5/2 + 1/2[\hat{\mu}_0(-(q + 1)\gamma_0) - \hat{\mu}_0(-q\gamma_0)]$$

(observe $(q + 2) \notin J$). Now $-(q - 1) \notin J$, so $-q \in J$ implies $-(q + 1) \in J$ yielding $\int_G f\, d\mu_0 = 5/2$, otherwise (that is, $-q \notin J$) $\int_G f\, d\mu_0 \geq 5/2$, hence $\| \mu_0 \| \geq 5/4$. □

3.6 Remark: With the same type of combinatorial argument, Saeki [2] has further proved that if μ is idempotent and $1 < \| \mu \| < ((17)^{1/2} + 1)/4$, then $\mu = (\gamma_1 + \gamma_2) \cdot m_H$, for some compact subgroup H, and γ_1, γ_2 distinct characters of H.

4. Remarks

4.1: One important application of the idempotent measures theorem is to homomorphisms of group algebras. The idea is this: Let G_1, G_2 be compact abelian groups, and ϕ a homomorphism of $L^1(G_1)$ into $M(G_2)$, then ϕ induces a map $\hat{\phi}$ from a subset Y of $\hat{G}_2$ into $\hat{G}_1$, such that $(\phi f)\hat{}\,(\gamma) = \hat{f}[\phi(\gamma)]$ for $\gamma \in Y$, 0 for $\gamma \notin Y$, and it can be shown that the graph of $\hat{\phi}$ is $S(\mu) = \{\gamma : \hat{\mu}(\gamma) = 1\}$ for some idempotent $\mu \in M(G_1 \oplus G_2)$. Thus $\hat{\phi}$ is a piecewise affine map. The theorem is then extended to LCA groups by using the Bohr compactifications. A full description and bibliography can be found in Rudin [R, Ch. 4]. The theorem for general LCA groups is due to Cohen [2].

Some work has been done in the nonabelian case dealing with isometric isomorphisms of measure algebras. Wendel [1] showed that for $L^1(G)$ an isometric isomorphism is possible essentially only in the form $\phi f(x) = \chi(x)$ $f(x)$, where χ is a homomorphism of G into **T**. Johnson [1] extended this result to the measure algebra. Strichartz [1] proved a similar result for algebras between $L^1(G)$ and $M(G)$.

4.2 Historical Notes: The important Theorem 2.6 is due to Cohen [1]. Lemmas 2.3, 2.4 and the proof of Theorem 2.6 are from the paper of Îto and Amemiya [1]. Sections 3.2, 3.3, and 3.4 are from Rudin [R, p. 62 and p. 73]. Theorem 3.5 was proved by Saeki [1].

CHAPTER 7

INTRODUCTION TO COMPACT GROUPS

1. Introduction

1.1: Without commutativity but with compactness for the group there is still a quite satisfactory theory of harmonic analysis. Matrix-valued coefficients take the place of scalar Fourier coefficients of an integrable function, although the series of matrices generally has arbitrarily large dimensions. The basic facts about representations of a compact group, such as the finite dimensionality of an irreducible unitary continuous representation, and the orthogonality of matrix entry functions of irreducible representations, are to be found in this chapter. Such properties of a representation, as weak or strong measurability, weak or strong continuity, complete decomposability, and their logical interconnections will also be studied here.

Henceforth, the groups under consideration will be not necessarily abelian.

1.2: Let G be a compact group. It has a unique left, right, and inverse invariant regular Borel measure, namely the Haar measure, m_G, normalized by $\int_G dm_G = 1$; $dm_G(x)$ will also be denoted by dx. (The construction of Haar measure is described in Appendix A.) If $f \in C(G)$ then $\int_G f(x)\,dx = \int_G f(xy)\,dx = \int_G f(yx)\,dx = \int_G f(x^{-1})\,dx$ for each $y \in G$. Define $L^p(G)$ to be $L^p(dm_G)$, and let G act on $L^p(G)$ by $R(x)f(y) = f(yx)$; then each $R(x)$ is an isometry of $L^p(G)$ onto itself. There is an analogous action on the left $f(y) \mapsto f(x^{-1}y)$. A function f on G is called **uniformly continuous** if for each $\varepsilon > 0$, there exists a neighborhood V of e (the identity in G) such that $|f(x) - f(y)| < \varepsilon$

whenever $x^{-1}y \in V$. Standard methods show that each $f \in C(G)$ is uniformly continuous, which means that for any $f \in C(G)$, $\| R(x)f - f \|_\infty \to 0$ as $x \to e$.

1.3 Proposition: *For each $f \in L^p(G)$, $1 \le p < \infty$ the map $G \to L^p(G)$ given by $x \mapsto R(x)f$ is continuous.*

Proof: Let $\varepsilon > 0$, then by the density of $C(G)$ in $L^p(G)$ there exists $g \in C(G)$ such that $\| f - g \|_p < \varepsilon/3$. There exists a neighborhood V of e such that $x \in V$ implies $\| g - R(x)g \|_\infty < \varepsilon/3$. Now for $x^{-1}y \in V$,

$$\begin{aligned} \| R(x)g - R(y)g \|_p &= \| g - R(x^{-1}y)g \|_p \\ &\le \| g - f \|_p + \| f - R(x^{-1}y)f \|_p \\ &\quad + \| R(x^{-1}y)f - R(x^{-1}y)g \|_p < \varepsilon. \quad \square \end{aligned}$$

1.4: Convolution is defined in the following theorem which is stated without proof since it is just like the corresponding theorem for abelian groups.

Theorem: (a) *If f, $g \in L^1(G)$ then the integral $f * g(x) = \int_G f(xy^{-1})g(y)\,dy$ is defined and finite for m_G-almost all $x \in G$ and $f * g \in L^1(G)$ with*

$$\| f * g \|_1 \le \| f \|_1 \, \| g \|_1.$$

(b) *If $f \in L^p(G)$, $g \in L^q(G)$, $1 \le p \le \infty$, $1/p + 1/q = 1$, then $f * g \in C(G)$ with $\| f * g \|_\infty \le \| f \|_p \, \| g \|_q$.*

(c) *If $f \in L^p(G)$, $g \in L^1(G)$, $1 \le p \le \infty$, then $f * g$ and $g * f$ are in $L^p(G)$ with $\| f * g \|_p$ and $\| g * f \|_p \le \| f \|_p \, \| g \|_1$.*

(d) *If $f \in L^p(G)$, $1 \le p \le \infty$, $\mu \in M(G)$ then*

$$f * \mu(x) = \int_G f(xy^{-1})\,d\mu(y)$$

and $\mu * f(x) = \int_G f(y^{-1}x)\,d\mu(y)$ define functions in $L^p(G)$, with $\| f * \mu \|_p$ and $\| \mu * f \|_p \le \| f \|_p \, \| \mu \|$.

(e) *If μ, $\nu \in M(G)$, then $\mu * \nu \in M(G)$ is defined by*

$$\int_G f\,d(\mu * \nu) = \int_G \int_G f(xy)\,d\mu(x)\,d\nu(y)$$

and

$$\| \mu * \nu \| \le \| \mu \| \, \| \nu \|;$$

this is consistent with (a) *if $f \in L^1(G)$ is interpreted as $f(x)\,dx$.*

(f) *Convolution is associative, and $L^1(G)$ is a Banach algebra.*

1.5 Definition: A **representation** of G on a normed linear space X is a map $T: G \to \mathcal{B}(X)$, the set of bounded linear operators on X, such that $T(xy) = T(x)T(y)$ for all x, $y \in G$ and $T(e) = I$, the identity. This implies that each

$T(x)$ has a bounded inverse. (The representation will sometimes be denoted by (T, X) when it is necessary to name the underlying space).

The representation is called

(a) **weakly measurable,** if $x \mapsto \phi(T(x)f)$ is m_G-measurable for all $f \in X$, $\phi \in X^*$, the dual of X.

(b) **weakly continuous** if $x \mapsto \phi(T(x)f)$ is in $C(G)$ for all $f \in X$, $\phi \in X^*$.

(c) **strongly measurable,** if $x \mapsto T(x)f$ is strongly m_G-measurable for each $f \in X$ (for a definition of strong measurability see [Y, p. 130]).

(d) **strongly continuous,** if $x \mapsto T(x)f$ is continuous for each $f \in X$.

(e) **unitary,** if X is a Hilbert space and $T(x^{-1}) = T(x)^*$.

(f) **cyclic,** if there exists $f \in X$ such that

$$\overline{Sp}\{T(x)f : x \in G\} = X.$$

Two representations of G, (T, V) and (S, W), are said to be **equivalent** if there exists an invertible linear map A of V onto W such that $AT(x) = S(x)A$ for all $x \in G$. Proposition 1.3 now gives the fact that $x \mapsto R(x)$ is a strongly continuous representation of G on each of $C(G)$ and $L^p(G)$, $1 \leq p < \infty$, and is unitary on $L^2(G)$.

If (T, X) is a representation of G then a subspace Y of X is said to be **invariant** if $T(x)Y \subset Y$ for all $x \in G$; then $T \mid Y$ is also a representation of G. If X has no proper nontrivial invariant subspaces then it is said to be **irreducible.** A very significant property of compact groups is that any weakly continuous unitary irreducible representation $(T, \mathscr{H})$ is finite dimensional (referring to the dimension of $\mathscr{H}$). Further, any weakly continuous unitary representation is strongly continuous and a direct sum of irreducible ones. An analogous theorem holds for strongly continuous representations on a Banach space (Section 4).

2. Theorems on Unitary Representations

2.1 Proposition: *If $(T, \mathscr{H})$ is a unitary representation of G and Y is an invariant subspace in $\mathscr{H}$, then $Y^\perp$ is also invariant. If Y is closed then T is the direct sum of $T \mid Y$ and $T \mid Y^\perp$.*

Proof: Let $f \in Y^\perp$ then

$$(g, U(x)f) = (U^*(x)g, f) = (U(x^{-1})g, f) = 0$$

for all $g \in Y$, thus $U(x)f \in Y^\perp$ for each $x \in G$. If Y is closed then $\mathscr{H} = Y \oplus Y^\perp$ and each $T(x)$ is reduced by Y. □

2.2 Proposition: *Any unitary representation $(T, \mathscr{H})$ is a direct sum of cyclic representations.*

Proof: For any $f \in \mathscr{H}$, $\mathscr{H}_1 = \overline{Sp}\{T(x)f : x \in G\}$ is a closed invariant subspace of $\mathscr{H}$. Then $\mathscr{H} = \mathscr{H}_1 \oplus \mathscr{H}_1^{\perp}$ and a Zorn's lemma argument applied to $\mathscr{H}_1^{\perp}$ finishes the proof. □

2.3 Theorem: *Any weakly continuous unitary representation $(T, \mathscr{H})$ is completely reducible, that is, $\mathscr{H}$ is a direct sum of invariant irreducible subspaces, and each of these is finite dimensional. The same conclusion holds if $(T, \mathscr{H})$ is weakly measurable and unitary and $\mathscr{H}$ is separable.*

Proof: By Proposition 2.2 we may assume that T is cyclic, that is, there exists $h \in \mathscr{H}$, $|h| = 1$ such that $\overline{Sp}\{T(x)h : x \in G\} = \mathscr{H}$. Define a sesquilinear form $[\cdot, \cdot]$ on $\mathscr{H}$ by

$$[f, g] = \int_G (f, T(x)h)(T(x)h, g)\, dx,$$

which is meaningful since $x \mapsto (f, T(x)h)$ is measurable and bounded. Then

(1) $|[f, g]| \leq \int_G |f|\, |T(x)h|^2\, |g|\, dx = |f|\, |g|$

(2) $[f, f] = \int_G |(f, T(x)h)|^2\, dx \geq 0$

(3) If $[f, f] = 0$ then $f = 0$, by one of the following arguments:

Trivially $f = 0$ if T is weakly continuous; otherwise $\mathscr{H}$ is separable, so let $\{f_n : n = 1, 2, \ldots\}$ be dense in $\mathscr{H}_1 = \overline{Sp}\{T(x)f : x \in G\}$. Observe that for each $g \in \mathscr{H}_1$, $(g, T(x)h) = 0$ (m_G) a.e., since each $T(x)f$ has this property and it is preserved under linear operations and pointwise limits. Now let $E_n \subset G$, with $m_G E_n = 0$ and $(f_n, T(x)h) = 0$ for $x \notin E_n$, and put $E = \bigcup_{n=1}^{\infty} E_n$. Then $m_G E = 0$ and $(g, T(x)h) = 0$ for all $g \in \mathscr{H}_1$ and $x \notin E$, thus

$$(T(y)f, T(x)h) = (f, T(y^{-1}x)h) = 0 \text{ for all } y \in G,\ x \notin E,$$

but then

$$(f, T(y)h) = 0 \text{ for all } y \in G \text{ so } f = 0.$$

Hence there exists a bounded positive definite operator H on $\mathscr{H}$ such that $[f, g] = (Hf, g)$ for all $f, g \in \mathscr{H}$. Further H commutes with each $T(x)$, for

$$\begin{aligned}(T(x)Hf, g) &= (Hf, T(x^{-1})g) \\ &= \int_G (f, T(y)h)(T(y)h, T(x^{-1})g)\, dy \\ &= \int_G (f, T(y)h)(T(xy)h, g)\, dy \\ &= \int_G (f, T(x^{-1}y)h)(T(y)h, g)\, dy \\ &= (HT(x)f, g) \text{ for all } f, g \in \mathscr{H}.\end{aligned}$$

Now H is compact, for let $\{f_n\}$ be a sequence in $\mathscr{H}$ with $|f_n| \leq 1$ then by the Eberlein-Šmulian theorem [DS, p. 430] there exists a weakly convergent subsequence of $\{f_n\}$, and we claim $\{Hf_n\}$ has a strongly convergent subsequence. By reindexing suppose $f_n \xrightarrow{n} f$ weakly, then

$$\begin{aligned}|H(f_n - f)|^2 &= \int_G (f_n - f, T(x)h)(T(x)h, H(f_n - f))\, dx \\ &= \int_G (f_n - f, T(x)h)\, dx \int_G (T(x)h, T(y)h)(T(y)h, f_n - f)\, dy\end{aligned}$$

and there is pointwise convergence to 0 in each integrand, and the whole integrand is bounded by $| f_n - f |^2 \leq 4$, so the dominated convergence theorem proves the claim.

The spectral theorem now shows that $H = \Sigma_n \lambda_n P_n$, where P_n is the projection on a finite dimensional subspace Y_n corresponding to the eigenvalue λ_n, and each $\lambda_n > 0$ since H is positive definite. Thus $H = \Sigma_n \oplus Y_n$, and each $T(x)$ commutes with P_n, so Y_n is invariant, for $n = 1, 2, \ldots$. The spaces Y_n are not necessarily irreducible, but since they are finite-dimensional they can be decomposed into irreducible subspaces. □

2.4: This theorem applied to $(R, L^2(G))$, the **right regular representation,** now gives a large supply of irreducible unitary representations of G, enough, in fact, to separate points in G since $L^2(G)$ does. Further any irreducible unitary continuous representation of G is equivalent to one appearing in the decomposition of $L^2(G)$, as will be seen by examining the matrix entries (as functions on G) of a representation. The matrix entry functions form an orthogonal basis for $L^2(G)$ and permit harmonic analysis with some similarity to that of a compact abelian group.

2.5 Definition: Let $\hat{G}$, the **dual of G,** be the set of all equivalence classes of continuous unitary irreducible representations of G, and if $\alpha \in \hat{G}$, let T_α be some element of the class α. Then T_α is a homomorphism of G into $\mathbf{U}(n_\alpha)$ (the group of $n_\alpha \times n_\alpha$ unitary matrices), and we use $T_\alpha(x)_{ij}$ to denote the matrix entries of $T_\alpha(x)$, $1 \leq i, j \leq n_\alpha$, and $T_{\alpha ij}$ to denote the function $x \mapsto T_\alpha(x)_{ij}$. Then

$$T_\alpha(xy)_{ij} = \Sigma_k\, T_\alpha(x)_{ik}\, T_\alpha(y)_{kj} \text{ and } T_\alpha(y^{-1})_{ij} = \overline{T_\alpha(y)_{ji}}, \text{ and } T_{\alpha ij} \in C(G).$$

2.6 Schur's Lemma: *Let (S, V) and (T, W) be irreducible representations of G and suppose A is a linear map: $V \to W$ such that $AS(x) = T(x)A$ for all $x \in G$, then either A is an isomorphism of V onto W or $A = 0$.*

Proof: Now ker A is an invariant subspace of V, for if $v \in \ker A$, then $AS(x)v = T(x)Av = 0$ for all x, thus ker $A = (0)$ or V. Also AV is invariant in W, for if $w \in AV$ then $w = Av$ for some $v \in V$ and $T(x)w = AS(x)v \in AV$, thus $AV = W$ or (0). □

2.7 Schur's Lemma: *Let (T, V) be an irreducible representation of G on a finite dimensional complex vector space V, and A a linear map: $V \to V$ such that $T(x)A = AT(x)$ for all $x \in G$, then $A = cI$ for some complex c.*

Proof: For any complex c, $(A - cI)T(x) = T(x)(A - cI)$ and by 2.6, $A - cI$ is either 0 or one-to-one. Then $A - cI = 0$ for some c, for otherwise det $(A - cI) \neq 0$ for all complex c. □

2.8 Theorem: *If $\alpha, \beta \in \hat{G}$, $\alpha \neq \beta$, then*

$$\int_G T_\alpha(x)_{ij}\overline{T_\beta(x)_{kl}}\,dx = 0, \qquad 1 \le i, j \le n_\alpha$$
$$1 \le k, l \le n_\beta.$$

Proof: Let C be a fixed $n_\alpha \times n_\beta$ complex matrix and

$$D = \int_G T_\alpha(x)CT_\beta(x^{-1})\,dx = \int_G T_\alpha(yx)CT_\beta(x^{-1}y^{-1})\,dx = T_\alpha(y)DT_\beta(y^{-1}),$$

that is, $T_\alpha(y)D = DT_\beta(y)$ for all $y \in G$. By 2.6, $D = 0$ since $\alpha \neq \beta$. Now put $C = E_{jl}$ (the matrix with entries $(E_{jl})_{rs} = \delta_{rj}\delta_{sl}$, $1 \le j \le n_\alpha$, $1 \le l \le n_\beta$), to obtain $0 = \int_G T_\alpha(x)E_{jl}T_\beta(x^{-1})\,dx$, that is, $\int_G T_\alpha(x)_{ij}T_\beta(x^{-1})_{lk}\,dx = 0$ for all i, k, but $T_\beta(x^{-1})_{lk} = \overline{T_\beta(x)_{kl}}$. □

2.9 Theorem: *Let $\alpha \in \hat{G}$, then*

$$\int_G T_\alpha(x)_{ij}\overline{T_\alpha(x)_{kl}}\,dx = \delta_{ik}\delta_{jl}/n_\alpha.$$

Proof: Let C be a fixed $n_\alpha \times n_\alpha$ matrix and $D = \int_G T_\alpha(x)CT_\alpha(x^{-1})\,dx$. As above $T_\alpha(x)D = DT_\alpha(x)$ so by 2.7, $D = \lambda I$ for some $\lambda \in \mathbf{C}$. Let Tr D denote $\Sigma_i D_{ii}$, the **trace** of D.

Then

$$\mathrm{Tr}\, D = \int_G \mathrm{Tr}\,(T_\alpha(x)CT_\alpha(x)^{-1})\,dx = \int_G (\mathrm{Tr}\, C)\,dx = \mathrm{Tr}\, C = n_\alpha\lambda;$$

hence $\lambda = \mathrm{Tr}\, C/n_\alpha$. Put $C = E_{jl}$, then

$$\begin{aligned}\int_G T_\alpha(x)_{ij}T_\alpha(x^{-1})_{lk}\,dx &= \int_G T_\alpha(x)_{ij}\overline{T_\alpha(x)_{kl}}\,dx \\ &= \delta_{ik}\,\mathrm{Tr}\,(E_{jl})/n_\alpha \\ &= \delta_{ik}\delta_{jl}/n_\alpha. \quad \square\end{aligned}$$

3. Harmonic Analysis of $L^2(G)$

3.1 Remark: For a fixed $\alpha \in \hat{G}$, fixed j, $1 \le j \le n_\alpha$, consider the n_α-dimensional space of functions $\{\Sigma_i c_i T_{\alpha ji}: c_i \in \mathbf{C}\}$. Then

$$\begin{aligned}R(x) \sum_i c_i T_\alpha(y)_{ji} &= \sum_i c_i T_\alpha(yx)_{ji} \\ &= \sum_i \sum_r c_i T_\alpha(y)_{jr} T_\alpha(x)_{ri} \\ &= \sum_i \Big(\sum_r T_\alpha(x)_{ir}c_r\Big)T_\alpha(y)_{ji};\end{aligned}$$

so R restricted to this space is equivalent to T_α.

3.2 Definition: For $\alpha \in G$, let $V_\alpha = Sp\{T_{\alpha ij}: 1 \le i, j \le n_\alpha\}$. (This is independent of the choice of T_α in α.)

3.3 Theorem: *$L^2(G) = \Sigma \oplus_{\alpha\in\hat{G}} V_\alpha$, and V_α decomposes (not necessarily uniquely) into n_α copies of T_α.*

Proof: From the orthogonality relations we have $V_\alpha \perp V_\beta$ for $\alpha \neq \beta$. By Theorem 2.3 applied to $(R, L^2(G))$ we have that $L^2(G) = \Sigma \oplus_{i\in J} \mathscr{H}_i$, where each $\mathscr{H}_i$ is irreducible, for some index set J. We claim that for each $\alpha \in \hat{G}$, there are exactly n_α spaces $\mathscr{H}_i$ in the class α, and $\Sigma\oplus\{\mathscr{H}_i : \mathscr{H}_i \in \alpha\} = V_\alpha$. In fact suppose $\mathscr{H}_i \in \alpha$, then there is a basis of n_α functions f_j in $\mathscr{H}_i$ such that $R(x)f_j = \sum_{k=1}^{n_\alpha} T_\alpha(x)_{kj} f_k$, $1 \leq j \leq n_\alpha$. Thus each f_j is continuous since

$$\begin{aligned}\int_G (f_j * n_\alpha \sum_i T_{\alpha ii})(x)\overline{h(x)}\,dx &= \int_G \sum_k n_\alpha \sum_i \overline{T_\alpha(y)_{ii}}\, T_\alpha(y)_{kj}\,dy \int_G f_k(x)\overline{h(x)}\,dx \\ &= \int_G f_j(x)\overline{h(x)}\,dx\end{aligned}$$

for all $h \in L^2(G)$ so $f_j = f_j * n_\alpha \Sigma_i T_{\alpha ii}$ in $L^2(G)$. Furthermore

$$f_j(x) = R(x)f_j(e) = \Sigma_k f_k(e)T_\alpha(x)_{kj},$$

implying that $\mathscr{H}_i = Sp\{f_j\} \subset V_\alpha$. The fact that $\dim V_\alpha = n_\alpha^2$ finishes the argument. □

3.4 Definition: For any $f \in L^2(G)$ let $n_\alpha \tilde{f}_\alpha$ be the projection of f on V_α. Observe that $f \sim \Sigma_{\alpha\in\hat{G}} n_\alpha \tilde{f}_\alpha$ (L^2-convergence). For $\alpha \in \hat{G}$, let $\chi_\alpha(x) = \mathrm{Tr}\,(T_\alpha(x)) = \Sigma_i T_\alpha(x)_{ii}$, then χ_α is called the **character** of α and is independent of the choice of T_α in α.

3.5 Proposition: $\tilde{f}_\alpha = f * \chi_\alpha$

Proof: We have

$$\begin{aligned} f * \chi_\alpha(x) &= \int_G f(xy^{-1})\chi_\alpha(y)\,dy \\ &= \int_G f(y) \sum_i T_\alpha(y^{-1}x)_{ii}\,dy \\ &= \int_G f(y) \sum_{i,j} T_\alpha(y^{-1})_{ij} T_\alpha(x)_{ji}\,dy \\ &= \sum_{i,j} \left(\int_G f(y)\overline{T_\alpha(y)_{ji}}\,dy\right) T_\alpha(x)_{ji},\end{aligned}$$

and

$$n_\alpha \tilde{f}_\alpha(x) = \sum_{i,j} \left(\int_G f(y)\sqrt{n_\alpha}\,\overline{T_\alpha(y)_{ji}}\,dy\right)\sqrt{n_\alpha}\,T_\alpha(x)_{ji},$$

(since $\| T_{\alpha ij} \|_2 = 1/\sqrt{n_\alpha}$). □

3.6 Theorem: *Characters have the following properties:*

(1) $\int_G \chi_\alpha(x)\overline{\chi_\beta(x)}\,dx = \delta_{\alpha\beta}$
(2) $\chi_\alpha * \chi_\beta = (\delta_{\alpha\beta}/n_\alpha)\chi_\alpha$
(3) *$\chi_\alpha(xy) = \chi_\alpha(yx)$ for all $x, y \in G$ so $f * \chi_\alpha = \chi_\alpha * f$ for all $f \in L^1(G)$*
(4) *If $f \in L^1(G)$ and $\mu \in M(G)$ then $f * \chi_\alpha$ and $\mu * \chi_\alpha$ are in V_α.*

3.7: There is also a matrix type expression for $\Sigma_\alpha n_\alpha \tilde{f}_\alpha \sim \Sigma_\alpha n_\alpha f * \chi_\alpha$. For $f \in L^1(G)$, $\alpha \in \hat{G}$, $1 \leq i, j \leq n_\alpha$, put

$$\hat{f}_{\alpha ij} = \int_G f(x) T_\alpha(x^{-1})_{ij}\, dx = \int_G f(x) \overline{T_\alpha(x)_{ji}}\, dx,$$

which in matrix notation is written $\hat{f}_\alpha = \int_G f(x) T_\alpha(x^{-1})\, dx$. Then

$$\tilde{f}_\alpha(x) = \sum_{i,j} \hat{f}_{\alpha ij} T_\alpha(x)_{ji} = \operatorname{Tr}(\hat{f}_\alpha T_\alpha(x)),$$

so if $f \in L^2(G)$, then $f \sim \Sigma_\alpha n_\alpha \operatorname{Tr}(f_\alpha T_\alpha(x))$. We also obtain the **Plancherel theorem**

$$\| f \|_2^2 = \sum_\alpha \sum_{i,j=1}^{n_\alpha} | \sqrt{n_\alpha} \hat{f}_{\alpha ij} |^2 = \sum_\alpha n_\alpha \operatorname{Tr}(\hat{f}_\alpha{}^* \hat{f}_\alpha).$$

This settles the harmonic analysis of $L^2(G)$. Another big question is the uniform approximation problem for $C(G)$, namely, is $Sp\{V_\alpha : \alpha \in \hat{G}\}$ (uniformly) dense in $C(G)$. We note here that finite linear combinations of $\{T_{\alpha ij} : \alpha \in \hat{G}\}$ are customarily called **trigonometric polynomials.**

3.8 Algebra of representations: Given two finite-dimensional continuous unitary representations (S, V) and (T, W), of dimensions m and n respectively, we can construct the **direct sum** $(S \oplus T, V \oplus W)$ of dimension $m + n$, and the **tensor product** $(S \otimes T, V \otimes W)$ of dimension mn. As for the characters of these representations, these operations correspond to ordinary addition and multiplication, respectively. In fact

$$\operatorname{Tr}(S(x) \oplus T(x)) = \operatorname{Tr}(S(x)) + \operatorname{Tr}(T(x))$$

and

$$\operatorname{Tr}(S(x) \otimes T(x)) = \operatorname{Tr}(S(x)) \cdot \operatorname{Tr}(T(x)).$$

A **conjugation** may be defined as follows: choose an orthonormal basis for V, and then (complex) conjugate the matrix entries of $S(x)$ with respect to this basis, yielding a representation $\overline{S}(x)$ (a different basis would result in a representation equivalent to $\overline{S}(x)$). Observe that

$$\operatorname{Tr}(\overline{S}(x)) = \overline{\operatorname{Tr} S(x)}.$$

Now any finite-dimensional continuous unitary representation T is a finite direct sum of irreducible ones, and the above statements indicate that the character of T carries enough information to find the irreducible components. In fact $T \cong \Sigma \oplus_\alpha m_\alpha T_\alpha$, where m_α is the number of times T_α appears in T, and $m_\alpha = \int_G \operatorname{Tr}(T(x)) \overline{\chi_\alpha(x)}\, dx$. This latter formula follows from $\operatorname{Tr}(T(x)) = \Sigma_\alpha m_\alpha \chi_\alpha(x)$ and the orthogonality relations 3.6. This incidentally yields the following **proposition:** T is irreducible if and only if $\int_G | \operatorname{Tr}(T(x)) |^2\, dx = 1$.

Conjugation applied to an irreducible representation of class α yields another irreducible representation, whose class is denoted by $\bar{\alpha}$, and of course $\chi_{\bar{\alpha}} = \overline{\chi_\alpha}$. A tensor product $T_\alpha \otimes T_\beta$ decomposes into irreducible components

$$T_\alpha \otimes T_\beta \cong \Sigma \oplus_\gamma m_{\alpha\beta}(\gamma) T_\gamma$$

where $m_{\alpha\beta}(\gamma) = \int_G \chi_\alpha \chi_\beta \overline{\chi_\gamma}\, dm_G$ (observe $m_{\alpha\beta}(\gamma) = m_{\alpha\bar{\gamma}}(\bar{\beta}) = m_{\bar{\gamma}\beta}(\bar{\alpha})$). If the group G is abelian then an application of Lemma 2.7 shows that any irreducible representation is one-dimensional, and the tensor product operation $T_\alpha \otimes T_\beta$ makes $\hat{G}$ into a discrete abelian group, the ordinary dual. Thus we use *dual* as a name for $\hat{G}$ in general.

3.9: The above considerations show that the trigonometric polynomials form a self-adjoint algebra of continuous functions, that is, if f, g are trigonometric polynomials then so are $\bar{f}$, fg, and $f + g$. This algebra separates points because $(R, L^2(G))$ does, so by the Stone-Weierstrass theorem the trigonometric polynomials are (uniformly) dense in $C(G)$. Another proof of this will be given in the next section as a special case of a general theorem on Banach representations.

4. Representations on a Banach Space

4.1: Suppose T is a representation of G on a Banach space X, then there is a skew representation T^* on X^*, and a representation T^{**} on X^{**}. For $x \in G$, $T^*(x)$ is defined by $(T^*(x)\phi)\,(f) = \phi(T(x)f)$ for all $\phi \in X^*$, $f \in X$ and $T^{**}(x)\,\Psi\,(\phi) = \Psi(T^*(x)\phi)$ for $\Psi \in X^{**}$, $\phi \in X^*$. Then for $x, y \in G$, $T^*(xy) = T^*(y)T^*(x)$ and $T^{**}(xy) = T^{**}(x)T^{**}(y)$.

4.2 Proposition: *If T is weakly continuous then*

$$M = \sup\{\, \| T(x) \| : x \in G\} < \infty.$$

Proof: For fixed $f \in X$, the collection $\{\phi \mapsto \phi(T(x)f) \colon x \in G\}$ of linear maps on X^*, is pointwise bounded, that is,

$$\sup\{\, |\phi(T(x)f)| : x \in G\} < \infty$$

for each ϕ, thus by the uniform boundedness principle $M_f = \sup\{\, \| T(x)f \| : x \in G\} < \infty$. Another application of this principle finishes the proof. □

4.3 Definition: The representation (T, X) is said to be **completely decomposable** if $\{X_j\}$, the set of finite dimensional invariant irreducible subspaces, spans a (strongly) dense subspace of X. Loosely speaking, this says that there are objects like trigonometric polynomials in X which are dense.

4.4 Proposition: *Let T be a weakly continuous representation of G on a Banach space X, then there exists a closed subspace $\tilde{X}$ of X^{**} such that T^{**} restricted to $\tilde{X}$ is strongly continuous and completely decomposable, and $\tilde{X}$ is total in X^{**}, that is, $\Phi(\phi) = 0$ for all $\Phi \in \tilde{X}$ implies $\phi = 0$ in X^*.*

Proof: Fix $f \in X$ and define a linear map $A_f : X^* \to L^2(G)$ by $A f \phi(x) = \phi(T(x^{-1})f)$ for $\phi \in X^*$. By Proposition 4.2, $\| A_f \phi \|_\infty \leq M \| \phi \| \| f \|$ for some constant M depending only on T, thus A_f is bounded. Furthermore, for each $x \in G$,

$$A_f(T^*(x)\phi)(y) = (T^*(x)\phi)(T(y^{-1})f) = \phi(T(xy^{-1})f)$$
$$= A_f\phi(yx^{-1}) = R(x^{-1})A_f\phi(y),$$

that is $A_f T^*(x) = R(x^{-1})A_f$. Let $\mathscr{H}_f$ be the L^2-closure of $A_f X^*$, then $\mathscr{H}_f$ is invariant, since $A_f X^*$ is and thus $\mathscr{H}_f = \Sigma \oplus_j \mathscr{H}_{f,j}$, where each $\mathscr{H}_{f,j}$ is an invariant irreducible finite-dimensional subspace. Now define A_f^*, a conjugate-linear map from $\mathscr{H}_f \to X^{**}$, by $A_f^* v(\phi) = \langle A_f \phi, v \rangle$ for $v \in \mathscr{H}_f$ (where $\langle , \rangle$ denotes the inner product in $L^2(G)$). This map is one-to-one since $A_f X^*$ is dense in $\mathscr{H}_f$, and $A_f^* R(x) = T^{**}(x) A_f^*$, for

$$A_f^*(R(x)v)(\phi) = \langle A_f\phi, R(x)v \rangle = \langle R(x^{-1})A_f\phi, v \rangle = \langle A_f T^*(x)\phi, v \rangle$$
$$= A_f^* v(T^*(x)\phi) = T^{**}(x)A_f^* v(\phi),$$

for $v \in \mathscr{H}_f$, $\phi \in X^*$. Thus each $A_f^* \mathscr{H}_{f,j}$ is an invariant irreducible finite dimensional subspace of X^{**} on which T^{**} is strongly continuous. Now let

$$\tilde{X} = \overline{Sp}\,\{A_f^* \mathscr{H}_{f,j} : f \in X,\ \mathscr{H}_f = \Sigma \oplus_j \mathscr{H}_{f,j}\},$$

then T^{**} is strongly continuous on $\tilde{X}$, and $\tilde{X}$ is obviously completely decomposable. Now suppose that $\phi \in X^*$ and $\Phi(\phi) = 0$ for all $\Phi \in \tilde{X}$. If $\phi \neq 0$ then there exists $f \in X$ such that $\phi(f) \neq 0$, but $A_f^* A_f \phi(\phi) = 0$, that is, $\int_G | \phi(T(x^{-1})f) |^2 \, dx = 0$, and by continuity $\phi(f) = 0$, a contradiction. □

4.5 Theorem: *Let T be a weakly continuous representation of G on a Banach space X, then the following are equivalent:*

(a) *T is completely decomposable*
(b) *T is strongly continuous*
(c) *for each $f \in X$, $\overline{Sp}\{T(x)f : x \in G\}$ is separable*
(d) *T is strongly measurable.*

Proof: (a) implies (b): for any $f \in X$, $\varepsilon > 0$, there exists $g \in X$ such that g is in an invariant finite-dimensional subspace, $x \mapsto T(x)g$ is continuous, and $\| f - g \| < \varepsilon/3\ (\sup_x \| T(x) \|)$. A triangle inequality argument shows that $x \mapsto T(x)f$ is continuous.

(b) implies (c): If T is strongly continuous and $f \in X$, then $E = \{T(x)f: x \in G\}$ is a compact subset of X, so E is a compact metric space, hence separable. Thus there exists a sequence $\{x_n\}$ in G such that $\{T(x_n)f\}$ is dense in E, and hence $\overline{Sp E}$ is separable.

(c) implies (d): By the Pettis theorem [Y, p. 131] a separably valued weakly measurable function is strongly measurable.

(d) implies (a): Proposition 4.4 provides a closed subspace $\tilde{X}$ of X^{**} which is completely decomposable, and it is enough to show $\tilde{X} \subset jX$, where j is the canonical map of X into X^{**}. since $\tilde{X}$ is already total. Further since $\tilde{X}$ is generated by $\{A_f^* A_f \phi: f \in X, \phi \in X^*\}$ it is enough to show each $A_f^* A_f \phi \in jX$ (this is because of the density of $\{A_f \phi\}$ in the domain of A_f^*).

For fixed $f \in X, \phi \in X^*$, define a function $g_{f,\phi}: G \rightarrow X$ by

$$g_{f,\phi}(x) = \overline{\phi(T(x^{-1})f)}T(x^{-1})f.$$

Then $g_{f,\phi}$ is strongly measurable by hypothesis, and

$$\| g_{f,\phi} \| \leq \| \phi \| \| f \|^2 \sup_x \| T(x) \|,$$

thus $g_{f,\phi}$ is Bochner integrable [Y, p. 133]. Now let $g'_{f,\phi} = \int_G g_{f,\phi}(x)\, dx$, then for $\psi \in X^*$,

$$\begin{aligned} A_f^* A_f \phi(\psi) &= \int_G \psi(T(x^{-1})f)\overline{\phi(T(x^{-1})f)}\, dx \\ &= \psi\left(\int_G \overline{\phi(T(x^{-1})f)}T(x^{-1})f\, dx\right) \\ &= \psi(g'_{f,\phi}), \text{ thus } A_f^* A_f \phi = jg'_{f,\phi}. \quad \square \end{aligned}$$

4.6 Corollary: *If X is separable or reflexive then the weak continuity of T implies strong continuity.*

4.7 Corollary: *The trigonometric polynomials are dense in $C(G)$, $L^p(G)$ for $1 \leq p < \infty$; and if G is also a Lie group, in $C^k(G)$, the functions which are k-times continuously differentiable, $k = 1, 2, \ldots$.*

4.8 Corollary: *If $\mu \in M(G)$ and $\int_G T_\alpha(x^{-1})\, d\mu(x) = 0$ for all $\alpha \in \hat{G}$ then $\mu = 0$.*

Proof: $\int_G f\, d\mu = 0$ for every trigonometric polynomial f, hence for every $f \in C(G)$, so $\mu = 0$. $\square$

5. Historical Notes

The theory of representations of a compact Lie group was developed by Peter and Weyl [1] in 1926. The construction of Haar measure for any compact group (Haar [1], Weil [W]) made it possible to extend their theory to any compact group. The presentation of Theorem 2.3 follows that of Shiga [1]. Section 4 is due to Shiga [1]. The above is not meant to be a listing of all works dealing with representations of compact groups.

CHAPTER 8

BANACH SPACES ON THE DUAL OF A COMPACT GROUP

1. Introduction

1.1: Throughout this chapter G is a **compact** group, and the notation and definitions of Chapter 7 are in force. As was seen in Chapter 7 the Fourier series of a function in $L^1(G)$ is a series of matrices (operators on finite dimensional spaces). In this chapter operator-valued analogues of the standard sequence spaces c_o, l^1, l^2, l^∞ will be studied and given group-theoretic significance. Section 2 presents various norms on operators, Section 3 contains definitions and theorems on operator series, while Section 4 relates them to the Fourier transform of G. The most important concept developed in this chapter is that of the Fourier algebra of G, $A(G)$, which is a Banach algebra of continuous functions on G, and is the analogue of $A(G) = (l^1(\hat{G}))\hat{}$ for a compact abelian group G. In fact, if G is a compact Lie group then only real-analytic functions operate in $A(G)$, just as for $A(\mathbf{T})$ ($\mathbf{T}$ = unit circle).

2. Norms on Finite Dimensional Linear Operators

2.1: In this section X and Y will denote finite dimensional complex inner-product spaces. Then $\mathscr{B}(X)$ is the algebra of linear maps $X \to X$ normed by the operator norm $\| \cdot \|_\infty$, that is, if $v \in \mathscr{B}(X)$, then

$$\| v \|_\infty = \sup\{ | v\xi | : \xi \in X, | \xi | \leq 1\}.$$

Then for $v, w \in \mathscr{B}(X)$ we have

$$\| v \|_\infty = \| v^* \|_\infty, \| vv^* \|_\infty = \| v \|_\infty^2,$$

and

$$\| vw \|_\infty \leq \| v \|_\infty \| w \|_\infty;$$

and if further v is unitary then $\| v \|_\infty = 1$. We define the **trace** Tr on $\mathscr{B}(X)$ by $\mathrm{Tr}(v) = \sum_{i=1}^n (v\xi_i, \xi_i)$ where $\{\xi_i\}_{i=1}^n$ is any orthonormal basis for X, then Tr is independent of the choice of $\{\xi_i\}$, $\mathrm{Tr}(uv) = \mathrm{Tr}(vu)$ for $u, v \in \mathscr{B}(X)$, and Tr is a linear functional. If ϕ is any linear functional on $\mathscr{B}(X)$ then there exists a unique $u \in \mathscr{B}(X)$ such that $\phi(v) = \mathrm{Tr}(uv)$, for all $v \in \mathscr{B}(X)$. We thus obtain another norm $\| \cdot \|_1$, dual to $\| \cdot \|_\infty$, on $\mathscr{B}(X)$, given by

$$\| u \|_1 = \sup\{ | \mathrm{Tr}(uv) | : \| v \|_\infty \leq 1\}.$$

If we denote the space $\mathscr{B}(X)$ with the norm $\| \cdot \|_1$ by $\mathscr{B}_1(X)$, then the dual space of $\mathscr{B}_1(X)$ is $\mathscr{B}(X)$, that is, with $\| \cdot \|_\infty$, and conversely (see 2.4). Observe that each $u \in \mathscr{B}(X)$ can be written as $u = w | u |$, where w is unitary, $| u |$ is a positive operator, and $| u |^2 = u^* u$.

2.2 Proposition: *For* $u \in \mathscr{B}(X)$, $\| u \|_1 = \mathrm{Tr}(| u |)$.

Proof: There exists an orthonormal basis $\{\xi_i\}_{i=1}^n$ for X of eigenvectors of $| u |$, so that the matrix entries of $| u |$ with respect to this basis are $| u |_{ij} = \lambda_i \delta_{ij}$, $\lambda_i \geq 0$. Then for $\| v \|_\infty \leq 1$ we have

$$\begin{aligned} | \mathrm{Tr}(uv) | &= | \mathrm{Tr}(w | u | v) | = | \mathrm{Tr}(| u | vw) | \\ &= \left| \sum_{i=1}^n \lambda_i (vw)_{ii} \right| \leq \sum_{i=1}^n \lambda_i = \mathrm{Tr}(| u |) \end{aligned}$$

since $| (vw)_{ii} | = | (vw\xi_i, \xi_i) | \leq \| vw \|_\infty = \| v \|_\infty \leq 1$. For $v = w^*$ equality is obtained in the above. □

2.3 Proposition: *Given* $\zeta, \eta \in X$ *define* $v \in \mathscr{B}(X)$ *by* $v\xi = (\xi, \zeta)\eta$, *then*

$$\| v \|_1 = \| v \|_\infty = | \zeta \| \eta | \text{ and } \mathrm{Tr}\, v = (\eta, \zeta).$$

2.4 Proposition: *For* $u \in \mathscr{B}(X)$

$$\| u \|_\infty = \sup\{ | \mathrm{Tr}(uv) | : \| v \|_1 \leq 1\} \text{ and}$$

$$\| u \|_\infty = \sup\{ \| uv \|_1 : \| v \|_1 \leq 1\}.$$

Proof: Suppose $u \neq 0$ then there exists $\xi_0 \in X$, $| \xi_0 | = 1$, such that $| u\xi_0 | = \| u \|_\infty$, then define v by

$$v \xi = (1/ \| u \|_\infty) (\xi, u\xi_0) \xi_0.$$

Then $\| v \|_1 = 1$ and $\| uv \|_1 = \mathrm{Tr}\,(uv) = \| u \|_\infty$. For any $v \in \mathscr{B}(X)$ it is clear that $|\mathrm{Tr}\,(uv)| \le \| u \|_\infty \| v \|_1$, and

$$\begin{aligned} \| uv \|_1 &= \sup\{ |\mathrm{Tr}\,(uvw)| : \| w \|_\infty \le 1\} \\ &\le \sup\{ |\mathrm{Tr}\,(vw)| : \| w \|_\infty \le \| u \|_\infty\} = \| u \|_\infty \| v \|_1. \quad \square \end{aligned}$$

2.5 Definition: There is an inner product on $\mathscr{B}(X)$, given by $\langle u, v\rangle = \mathrm{Tr}\,(v^* u)$. Then $\langle u, u\rangle = \mathrm{Tr}\,(u^* u) = \sum_{i,j=1}^n |u_{ij}|^2$, thus $\| u \|_2 = \langle u, u\rangle^{1/2}$ is a norm. Observe that $\| u^* \|_2 = \| u \|_2$.

2.6 Proposition: *For $u \in \mathscr{B}(X)$, $\| u \|_\infty \le \| u \|_2 \le \| u \|_1$.*

Proof: Let $\{\lambda_i\}_{i=1}^n$ be the set of eigenvalues of $|u|$, then $\| u \|_\infty = \max \lambda_i$, $\| u \|_2 = (\Sigma_i \lambda_i^2)^{1/2}$, and $\| u \|_1 = \Sigma_i \lambda_i$. $\square$

2.7 Proposition: *For $u, v \in \mathscr{B}(X)$,*

$$\| uv \|_2 \le \| u \|_\infty \| v \|_2,$$

and

$$\| u \|_\infty = \sup\{ \| uv \|_2 : \| v \|_2 \le 1\}.$$

Proof: We have

$$\begin{aligned} \| uv \|_2^2 &= \mathrm{Tr}\,(uvv^* u^*) = \mathrm{Tr}\,(u^* uvv^*) \le \| u^* u \|_\infty \mathrm{Tr}\,(|vv^*|) \\ &= \| u \|_\infty^2 \| v \|_2^2. \end{aligned}$$

The other conclusion follows from the same proof as 2.4, by using 2.6. $\square$

2.8 Proposition: *If $u, v \in \mathscr{B}(X)$, then $\| uv \|_1 \le \| u \|_2 \| v \|_2$ and for each $z \in \mathscr{B}(X)$ there exists u, v such that $z = uv$ and $\| z \|_1 = \| u \|_2 \| v \|_2$.*

Proof: Given $u, v \in \mathscr{B}(X)$ there exists a unitary $w \in \mathscr{B}(X)$ such that $\| uv \|_1 = \mathrm{Tr}\,(wuv)$, so

$$\| uv \|_1 = \langle (wu)^*, v\rangle \le \mathrm{Tr}\,(wuu^* w^*)^{1/2}\, \mathrm{Tr}\,(v^* v)^{1/2} = \| u \|_2 \| v \|_2$$

(the Cauchy–Schwarz inequality). Now suppose $z \in \mathscr{B}(X)$ then $z = w|z|$ with w unitary. Let $u = w|z|^{1/2}$, $v = |z|^{1/2}$ where $|z|^{1/2}$ exists since $|z|$ is positive. Then $z = uv$ and $\| u \|_2 = \mathrm{Tr}\,(|z|^{1/2*} w^* w |z|^{1/2})^{1/2} = \mathrm{Tr}\,(|z|)^{1/2}$ and $\| v \|_2 = \mathrm{Tr}\,(|z|)^{1/2}$ so $\| z \|_1 = \| u \|_2 \| v \|_2$. $\square$

2.9 Proposition: *Suppose $u \in \mathscr{B}(X)$, $v \in \mathscr{B}(Y)$, then*

$$u \otimes v \in \mathscr{B}(X \otimes Y)$$

and

$$\| u \otimes v \|_1 = \| u \|_1 \| v \|_1.$$

Proof: We have

$$\| u \otimes v \|_1 = \mathrm{Tr}(|u \otimes v|) = \mathrm{Tr}(|u| \otimes |v|)$$
$$= \mathrm{Tr}(|u|)\,\mathrm{Tr}(|v|) = \| u \|_1 \| v \|_1. \quad \square$$

2.10 Proposition: *Suppose $\{P_i\}_{i=1}^m$ is a set of mutually orthogonal projections on X such that $\sum_{i=1}^m P_i = I$ (the identity), that is, $P_i^* = P_i$, $P_iP_j = 0$ for $i \neq j$, and $X = \Sigma \oplus_{i=1}^m P_iX$. Then for any $u \in \mathscr{B}(X)$, $\sum_{i=1}^m \| P_iuP_i \|_1 \leq \leq \| u \|_1$.*

Proof: Let $v_i \in \mathscr{B}(P_iX)$ with $\| v_i \|_\infty \leq 1$ for each $i = 1, \ldots, m$, then

$$\textstyle\sum_{i=1}^m P_iv_iP_i \in \mathscr{B}(X),$$

where range $v_i \subset P_iX \subset X$, and

$$\Big\| \sum_{i=1}^m P_iv_iP_i \Big\|_\infty \leq 1.$$

Then

$$\| u \|_1 \geq \left| \mathrm{Tr}\left(u \sum_{i=1}^m P_iv_iP_i \right) \right| = \left| \mathrm{Tr}\left(\sum_{i=1}^m P_iuP_iv_i \right) \right|,$$

and now taking the supremum over all $\{v_i\}_{i=1}^m$, $\| v_i \|_\infty \leq 1$, we obtain the required inequality. $\square$

3. Generalized Sequence Spaces

3.1: For each $\alpha \in \hat{G}$ (see 7.2.5) we choose $(T_\alpha, X_\alpha) \in \alpha$ (that is, T_α is a continuous unitary irreducible representation of G on an n_α-dimensional vector space X_α).

3.2 Definition: Let ϕ be a set $\{\phi_\alpha : \alpha \in \hat{G}, \phi_\alpha \in \mathscr{B}(X_\alpha)\}$ such that

$$\sup_\alpha \| \phi_\alpha \|_\infty < \infty.$$

The set of all such ϕ will be denoted by $\mathscr{L}^\infty(\hat{G})$ (or $\mathscr{L}^\infty$ for short) and is clearly a Banach space with the norm $\| \phi \|_\infty = \sup_\alpha \| \phi_\alpha \|_\infty$ and coordinatewise operations. Further $\mathscr{L}^\infty$ is a $*$-algebra with multiplication defined by $(\phi\psi)_\alpha = \phi_\alpha\psi_\alpha$, $\alpha \in \hat{G}$, and $(\phi^*)_\alpha = (\phi_\alpha)^*$. Further $\| \phi * \phi \|_\infty = \| \phi \|_\infty^2$ so $\mathscr{L}^\infty$ is a C^*-algebra. There is an identity $I = \{I \cdot_{X_\alpha} : \alpha \in \hat{G}\}$.

3.3 Definition: For $\phi \in \mathscr{L}^\infty$, put

$$\| \phi \|_1 = \sum_\alpha n_\alpha \| \phi_\alpha \|_1, \qquad \| \phi \|_2 = \Big(\sum_\alpha n_\alpha \| \phi_\alpha \|_2^2 \Big)^{1/2},$$

and let

$$\mathscr{L}^1(\hat{G}) = \{\phi\colon \| \phi \|_1 < \infty\}, \qquad \mathscr{L}^2(\hat{G}) = \{\phi\colon \| \phi \|_2 < \infty\}.$$

Then $\mathscr{L}^1$ is a Banach space, and $\mathscr{L}^2$ is a Hilbert space with inner product

$$\langle \phi, \psi \rangle = \sum_{\alpha} n_\alpha \operatorname{Tr}(\psi_\alpha{}^* \phi_\alpha).$$

3.4 Definition: Recall that for each $\alpha \in \hat{G}$ there exists a conjugate class $\bar{\alpha}$, and there exists a conjugate linear isometry $J\colon \Sigma\oplus_\alpha X_\alpha \to \Sigma\oplus_\alpha X_\alpha$ such that $J^2 = I$, $JX_\alpha = X_{\bar{\alpha}}$ and $JT_\alpha(x) = T_{\bar{\alpha}}(x)J$.

3.5 Proposition: *The map $\phi \mapsto J\phi J$ is a conjugate-linear isometric mapping of $\mathscr{L}^p$ onto itself for $p = 1, 2, \infty$. Observe $(J\phi J)_\alpha = J\phi_{\bar{\alpha}} J$.*

3.6 Proposition: *If $p = 1, 2$, $\phi \in \mathscr{L}^p$, $\psi \in \mathscr{L}^\infty$ then*

$$\psi\phi \in \mathscr{L}^p \quad \text{and} \quad \| \psi\phi \|_p \le \| \psi \|_\infty \| \phi \|_p.$$

Furthermore

$$\| \psi \|_\infty = \sup\{ \| \psi\phi \|_p\colon \| \phi \|_p \le 1\}.$$

Proof: Since $\| \psi_\alpha\phi_\alpha \|_p \le \| \psi_\alpha \|_\infty \| \phi_\alpha \|_p$ for each $\alpha \in \hat{G}$ (2.4 and 2.7) we obtain $\| \psi\phi \|_p \le \| \psi \|_\infty \| \phi \|_p$. Now let $\psi \in \mathscr{L}^\infty$, $\psi \ne 0$, $\varepsilon > 0$, then there exists $\alpha \in \hat{G}$ such that $\| \psi_\alpha \|_\infty > \| \psi \|_\infty - \varepsilon$. By Propositions 2.3, 2.4, and 2.7 there exists $\phi'_\alpha \in \mathscr{B}(X_\alpha)$ such that

$$\| \phi'_\alpha \|_1 = \| \phi'_\alpha \|_2 = 1$$

and

$$\| \psi_\alpha\phi'_\alpha \|_1 = \| \psi_\alpha\phi'_\alpha \|_2 = \| \psi_\alpha \|_\infty.$$

For $p = 1$, put $\phi = \{\phi_\beta\colon \beta \in \hat{G},\ \phi_\beta = \delta_{\alpha\beta}\phi'_\alpha/n_\alpha\}$, then

$$\| \phi \|_1 = \| \phi'_\alpha \|_1 = 1$$

and

$$\| \psi\phi \|_1 = \| \psi_\alpha \|_\infty > \| \psi \|_\infty - \varepsilon.$$

For $p = 2$, replace ϕ'_α/n_α by $\phi'_\alpha/\sqrt{n_\alpha}$. □

This exhibits $\mathscr{L}^\infty$ as a $*$-algebra of operators on a Hilbert space $\mathscr{L}^2$.

3.7 Definition: Let $\mathscr{C}_0(\hat{G}) = \{\phi \in \mathscr{L}^\infty\colon \lim_{\alpha\to\infty} \| \phi_\alpha \|_\infty = 0\}$, that is, for $\phi \in \mathscr{C}_0$ and for each $\varepsilon > 0$ there exists a finite set $S \subset \hat{G}$ such that

$$\| \phi_\alpha \|_\infty < \varepsilon$$

for all $\alpha \notin S$. $\mathscr{C}_0$ is a closed $*$-ideal in $\mathscr{L}^\infty$.

Let $\mathscr{C}_F(\hat{G}) = \{\phi \in \mathscr{L}^\infty\colon \phi_\alpha = 0$ for all but a finite number of $\alpha \in \hat{G}\}$.

Then $\mathscr{C}_F$ is a $*$-subalgebra of $\mathscr{C}_0$ and $\mathscr{C}_F \subset \mathscr{L}^1 \subset \mathscr{L}^2 \subset \mathscr{C}_0$, since

$$\| \phi_\alpha \|_\infty \leq \| \phi_\alpha \|_2 \leq \| \phi_\alpha \|_1$$

for any ϕ, $\alpha \in \hat{G}$ (see 2.6). $\mathscr{C}_F$ is dense in each of $\mathscr{L}^1$, $\mathscr{L}^2$, $\mathscr{C}_0$.

3.8 Definition: For $\phi \in \mathscr{L}^1$ define $\mathrm{Tr}\, \phi = \Sigma_\alpha n_\alpha \mathrm{Tr}'(\phi_\alpha)$. This is valid since $\Sigma_\alpha n_\alpha |\mathrm{Tr}\,(\phi_\alpha)| \leq \Sigma_\alpha n_\alpha \| \phi_\alpha \|_1 < \infty$.

3.9 Proposition: *There is a pairing of $\mathscr{L}^\infty$ and $\mathscr{L}^1$ given by $(\phi, \psi) \mapsto \mathrm{Tr}\,(\phi\psi)$ for $\phi \in \mathscr{L}^\infty$, $\psi \in \mathscr{L}^1$, and with this pairing $\mathscr{L}^1$ is identified with the dual of $\mathscr{C}_0$, and $\mathscr{L}^\infty$ with the dual of $\mathscr{L}^1$.*

Proof: By 3.8,

$$|\mathrm{Tr}\,(\phi\psi)| \leq \| \phi\psi \|_1 \leq \| \phi \|_\infty \| \psi \|_1.$$

Let Ψ be a bounded linear functional on $\mathscr{C}_0$, then by 2.1, there exists

$$\psi = \{\psi_\alpha : \alpha \in \hat{G}, \psi_\alpha \in \mathscr{B}(X_\alpha)\}$$

such that $\Psi(\phi) = \mathrm{Tr}\,(\phi\psi)$ for each $\phi \in \mathscr{C}_F$. Furthermore

$$|\mathrm{Tr}\,(\phi\psi)| = |\sum_\alpha n_\alpha \mathrm{Tr}\,(\phi_\alpha\psi_\alpha)| \leq \| \Psi \| \| \phi \|_\infty$$

for each $\phi \in \mathscr{C}_F$, so $\Sigma_\alpha n_\alpha \| \psi_\alpha \|_1 \leq \| \Psi \|$. Since $\mathscr{C}_F$ is dense in $\mathscr{C}_0$ this shows that $\Psi(\phi) = \mathrm{Tr}\,(\phi\psi)$ for all $\phi \in \mathscr{C}_0$. A similar argument using 2.4 shows that $\mathscr{L}^\infty$ is the dual of $\mathscr{L}^1$. □

3.10 Proposition: *If $\phi, \psi \in \mathscr{L}^2$ then $\phi\psi \in \mathscr{L}^1$ and*

$$\| \phi\psi \|_1 \leq \| \phi \|_2 \| \psi \|_2.$$

Further for any $\omega \in \mathscr{L}^1$ there exist $\phi, \psi \in \mathscr{L}^2$ such that

$$\omega = \phi\psi$$

and

$$\| \omega \|_1 = \| \phi \|_2 \| \psi \|_2.$$

Proof: We have

$$\begin{aligned} \| \phi\psi \|_1 &= \sum_\alpha n_\alpha \| \phi_\alpha\psi_\alpha \|_1 \\ &\leq \sum_\alpha n_\alpha \| \phi_\alpha \|_2 \| \psi_\alpha \|_2 \text{ (by 2.8)} \\ &\leq (\sum_\alpha n_\alpha \| \phi_\alpha \|_2^2)^{1/2} (\sum_\alpha n_\alpha \| \psi_\alpha \|_2^2)^{1/2} \\ &= \| \phi \|_2 \| \psi \|_2. \end{aligned}$$

Further given $\omega \in \mathscr{L}^1$, we construct for each $\alpha \in \hat{G}$, $\phi_\alpha, \psi_\alpha \in \mathscr{B}(X_\alpha)$ such that $\omega_\alpha = \phi_\alpha \psi_\alpha$ and $\| \phi_\alpha \|_2 = \| \psi_\alpha \|_2 = \| \omega_\alpha \|_1^{1/2}$ (by 2.8). Then

$$\| \phi \|_2^2 = \sum_\alpha n_\alpha \| \phi_\alpha \|_2^2 = \sum_\alpha n_\alpha \| \omega_\alpha \|_1 = \| \omega \|_1$$

and similarly for ψ. □

4. The Fourier Transform

4.1: In this section we show that the Fourier transform maps $L^1(G)$ one-to-one onto a dense subspace of $\mathscr{C}_0(\hat{G})$, $L^2(G)$ isometrically and isomorphically onto $\mathscr{L}^2(\hat{G})$, and the inverse transform maps $\mathscr{L}^1(\hat{G})$ onto a subalgebra of $C(G)$, which will be denoted by $A(G)$, the **Fourier algebra** of G.

4.2 Proposition: *The Fourier transform $\mathscr{F}$ on $L^1(G)$ defined by*

$$(\mathscr{F} f)_\alpha = \hat{f}_\alpha = \int_G f(x) T_\alpha(x^{-1})\, dx \text{ for } \alpha \in \hat{G}$$

is a bounded linear map into $\mathscr{L}^\infty(\hat{G})$ and has the following properties:

(i) $\| \hat{f} \|_\infty \leq \| f \|_1$
(ii) $\mathscr{F}(f^*) = (\mathscr{F} f)^*$, *where* $f^*(x) = \overline{f(x^{-1})}$, *for all* $x \in G$.
(iii) *If* $f, g \in L^1(G)$, *then* $\mathscr{F}(f * g) = (\mathscr{F} g)(\mathscr{F} f)$
(iv) $\mathscr{F}(\bar{f}) = J(\mathscr{F} f) J$
(v) $\mathscr{F}$ *is one-to-one.*

Proof: Let $f \in L^1(G)$, $\alpha \in \hat{G}$, $\xi, \eta \in X_\alpha$ then

$$|(\hat{f}_\alpha \xi, \eta)| = |\int_G f(x)(T_\alpha(x^{-1})\xi, \eta)\, dx| \leq \| f \|_1 |\xi \| \eta|,$$

thus $\| \hat{f}_\alpha \|_\infty \leq \| f \|_1$ for each α, proving (i). Parts (ii) and (iii) are easy computations, and part (v) is Corollary 7.4.8. Finally

$$(\mathscr{F}\bar{f})_\alpha = \int_G \overline{f(x)} T_\alpha(x^{-1})\, dx = \int_G \overline{f(x)} J T_{\bar{\alpha}}(x^{-1}) J\, dx$$
$$= \int_G J f(x) T_{\bar{\alpha}}(x^{-1}) J\, dx = J(\mathscr{F} f)_{\bar{\alpha}} J. \quad \square$$

4.3 Proposition: *The map $\mathscr{F}$ restricted to $L^2(G)$ is an isometric isomorphism onto $\mathscr{L}^2(\hat{G})$.*

Proof: The Plancherel theorem stated in 7.3.7 and the Riesz-Fischer theorem prove the proposition. □

4.4 Proposition: *The Fourier transform $\mathscr{F}$ defined on $M(G)$ by*

$$(\mathscr{F}\mu)_\alpha = \hat{\mu}_\alpha = \int_G T_\alpha(x^{-1})\, d\mu(x),\ \mu \in M(G),\ \alpha \in \hat{G}$$

is a linear map into $\mathscr{L}^\infty(\hat{G})$, and has the following properties:

(i) $\| \hat{\mu} \|_\infty \leq \| \mu \|$

(ii) $\mathscr{F}(\mu^*) = (\mathscr{F}\mu)^*$, *where* $\mu^*(E) = \overline{\mu(E^{-1})}$, *E Borel.*

(iii) *If* $\mu, v \in M(G)$ *then* $\mathscr{F}(\mu * v) = (\mathscr{F}v)(\mathscr{F}\mu)$.

(iv) $\mathscr{F}$ *is one-to-one.*

The proof is analogous to that of 4.2.

4.5 Proposition: *Each $f \in L^1(G)$ defines a bounded operator $R(f)$ on $L^2(G)$ by $R(f)g = g * f$, for $g \in L^2(G)$. Similarly given $\mu \in M(G)$ there is an operator $R(\mu)$, where $R(\mu)g = g * \mu$. Then*

$$R(f^*) = R(f)^*, \qquad R(\mu^*) = R(\mu)^*$$

and

$$\| R(f) \| = \| \hat{f} \|_\infty, \qquad \| R(\mu) \| = \| \hat{\mu} \|_\infty.$$

Proof: It is easy to check that $R(f^*) = R(f)^*$. Observe that $\mathscr{F}(R(f)g) = \hat{f}\hat{g}$ and Proposition 3.6 shows that

$$\| \hat{f} \|_\infty = \sup \{ \| \hat{f}\hat{g} \|_2 : \| \hat{g} \|_2 = \| g \|_2 \leq 1 \}. \quad \square$$

This proposition gives another interpretation of $\| \hat{f} \|_\infty$.

4.6 Proposition: *$\mathscr{F}$ maps the trigonometric polynomials onto $\mathscr{C}_F(\hat{G})$, and if f is a trigonometric polynomial then*

$$f(x) = \sum_\alpha n_\alpha \operatorname{Tr}(T_\alpha(x)\hat{f}_\alpha)$$

(a finite sum).

Proof: This follows from 7.3.5 and 7.3.7. $\square$

4.7 Proposition: *$\mathscr{F}$ maps $L^1(G)$ onto a dense $*$-subalgebra of $\mathscr{C}_0(\hat{G})$.*

Proof: $\mathscr{F}L^1(G)$ is a $*$-algebra in $\mathscr{L}^\infty$ by 4.2. Let $f \in L^1(G)$ then there is a sequence $\{f_n\} \subset L^2(G)$ such that $\| f_n - f \|_1 \xrightarrow{n} 0$. Then

$$\| \hat{f}_n - \hat{f} \|_\infty \leq \| f_n - f \|_1 \text{ so } \hat{f}_n \xrightarrow{n} \hat{f} \text{ in } \mathscr{L}^\infty,$$

but $f_n \in \mathscr{L}^2 \subset \mathscr{C}_0$, thus $f \in \mathscr{C}_0$. Finally $\mathscr{C}_F \subset \mathscr{F}L^1(G)$ and $\mathscr{C}_F$ is dense in $\mathscr{C}_0$. $\square$

4.8 Definition: For each $x \in G$, let $\hat{R}(x) \in \mathscr{L}^\infty(\hat{G})$ be defined by

$$\hat{R}(x)_\alpha = T_\alpha(x), \qquad \alpha \in \hat{G}.$$

Then for each $f \in L^2(G)$ we have $\mathscr{F}(R(x)f) = \hat{R}(x)\,\mathscr{F} f$, since

$$(R(x)f)\hat{}_\alpha = T_\alpha(x)\hat{f}_\alpha.$$

Thus $(\hat{R}, \mathscr{L}^2(\hat{G}))$ is a unitary representation equivalent to $(R, L^2(G))$ and is strongly continuous.

4.9 Proposition: *For a fixed $\phi \in \mathscr{L}^1(G)$ the map $x \mapsto \hat{R}(x)\phi$ is uniformly continuous on G, that is, $(\hat{R}, \mathscr{L}^1(\hat{G}))$ is strongly continuous.*

Proof: By Proposition 3.10 there exist $\psi, \omega \in \mathscr{L}^2$ such that $\phi = \psi\omega$. Then

$$\hat{R}(x)\phi - \hat{R}(y)\phi = (\hat{R}(x)\psi - \hat{R}(x)\psi)\omega$$

so

$$\| \hat{R}(x)\phi - \hat{R}(y)\phi \|_1 \leq \| \hat{R}(x)\psi - \hat{R}(y)\psi \|_2 \, \| \omega \|_2,$$

and the map $x \mapsto \hat{R}(x)\phi$ is uniformly continuous. □

4.10 Definition: The **inverse Fourier transform** $\mathscr{F}^{-1}$ is defined by

$$\mathscr{F}^{-1}\phi(x) = \operatorname{Tr}(\hat{R}(x)\phi)$$

for $\phi \in \mathscr{L}^1(\hat{G})$, $x \in G$. This transform is the unique extension to $\mathscr{L}^1$ of the inversion formula for trigonometric polynomials given in 4.6, namely $f(x) = \operatorname{Tr}(\hat{R}(x)\hat{f})$.

4.11 Proposition: *The inverse Fourier transform has the following properties:*

(i) *If f is a trigonometric polynomial, then $\mathscr{F}^{-1}(\mathscr{F} f) = f$.*
(ii) *$\mathscr{F}^{-1}$ is a bounded linear mapping of $\mathscr{L}^1(\hat{G})$ onto a dense subspace of $C(G)$, and $\| \mathscr{F}^{-1}\phi \|_\infty \leq \| \phi \|_1$, for all $\phi \in \mathscr{L}^1(\hat{G})$.*
(iii) *$\mathscr{F}^{-1}(\hat{R}(x)\phi) = R(x)\,\mathscr{F}^{-1}\phi$, for all $\phi \in \mathscr{L}^1$, $x \in G$.*
(iv) *$\mathscr{F}(\mathscr{F}^{-1}\phi) = \phi$ for each $\phi \in \mathscr{L}^1$, so $\mathscr{F}^{-1}$ is one-to-one.*

Proof: Proposition 4.6 gives (i). Given $x \in G$, $\phi \in \mathscr{L}^1$ we have $| \mathscr{F}^{-1}\phi(x) | \leq \| \phi \|_1$, and since $\mathscr{C}_F$ is dense in $\mathscr{L}^1$ we have that $\mathscr{F}^{-1}\phi$ is a continuous function for each $\phi \in \mathscr{L}^1$. Since the trigonometric polynomials are in $\mathscr{F}^{-1}\,\mathscr{L}^1$ we have that $\mathscr{F}^{-1}\,\mathscr{L}^1$ is dense in $C(G)$. Finally Fubini's theorem and the orthogonality relations show that $\mathscr{F}(\mathscr{F}^{-1}\phi) = \phi$. □

4.12 Definition: Denote $\mathscr{F}^{-1}\,\mathscr{L}^1(\hat{G})$ by $A(G)$. Then $A(G)$ consists of exactly those continuous functions f for which $\| f \|_A = \| \mathscr{F} f \|_1 = \Sigma_\alpha n_\alpha \| \hat{f}_\alpha \|_1$ is finite, that is, $\mathscr{F} f \in \mathscr{L}^1$. Further $A(G)$ is a Banach space under the norm $\| \cdot \|_A$. Since $\mathscr{L}^1$ is closed under the conjugation $\phi \to J\,\phi\,J$ (see 3.5), $A(G)$ is closed under conjugation $f \to \bar{f}$ and $\| \bar{f} \|_A = \| f \|_A$ (see 4.2 (iv)). Proposition 4.9 and 4.11 (iii) show that $A(G)$ is invariant under $R(x)$ and $(R, A(G))$ is a strongly continuous representation of G.

4.13 Corollary: *If $g, h \in L^2(G)$ then*

$$g * h \in A(G) \text{ and } \| g * h \|_A \leq \| g \|_2 \| h \|_2,$$

and if $f \in A(G)$ then there exist $g, h \in L^2(G)$ such that

$$f = g * h \text{ and } \| f \|_A = \| g \|_2 \| h \|_2.$$

Proof: This follows directly from 3.10, with $\omega = \mathscr{F} f$, $\mathscr{F} g = \psi$, $\mathscr{F} h = \phi$. □

4.14 Proposition: *Let $f \in C(G)$ then*

$$\| f \|_A = \sup\{ | \textstyle\int_G f(x)g(x)\, dx | : g \in L^1(G), \| \hat{g} \|_\infty \leq 1\}.$$

Proof: Suppose there is a constant $M < \infty$ such that

$$| \textstyle\int_G f(x)g(x)\, dx | \leq M \| \hat{g} \|_\infty$$

for all $g \in L^1$, then the map $g \mapsto \int fg$ extends uniquely to a bounded linear functional on $\mathscr{C}_0(\hat{G})$ (by 4.7). Thus by 3.9, $f = \mathscr{F}^{-1}\phi$ for some $\phi \in \mathscr{L}^1$, and

$$\| \hat{f} \|_1 = \sup\{ | \textstyle\int_G fg\, dm_G | : g \in L^1, \| \hat{g} \|_\infty \leq 1\}.$$

Observe that $\int_G fg dm_G = \text{Tr}\,(\hat{h}\hat{g})$ where $h(x) = f(x^{-1})$ and $\hat{h} = J\hat{f}^* J$, so $\| \hat{h} \|_1 = \| \hat{f} \|_1$. □

4.15 Proposition: *Let (T, X) be a continuous unitary finite-dimensional representation of G, let $u \in \mathscr{B}(X)$, and let $f(x) = \text{Tr}\,(T(x)u)$ for all $x \in G$, then $f \in A(G)$ and $\| f \|^A \leq \| u \|_1$.*

Proof: Observe that if T is irreducible then $T \cong T_\alpha$ for some α and

$$\| f \|_A = n_\alpha \| (1/n_\alpha)\, u \|_1 = \| u \|_1$$

(by the unitary invariance of Tr and $\| \cdot \|_1$). Otherwise T is a direct sum of irreducible representations, so $T \cong \Sigma \oplus_\alpha m_\alpha T_\alpha$ as in 7.3.8. Thus there is a set of mutually orthogonal projections $\{P_i\}_{i=1}^m$ on X such that $\sum_{i=1}^m P_i = I$ and T restricted to $P_i X$ is equivalent to some T_α. Thus $T(x) = \sum_{i=1}^m P_i T(x) P_i$ and

$$\text{Tr}\,(uT(x)) = \text{Tr}\left(\sum_{i=1}^m P_i u P_i T(x) P_i \right)$$

$$= \sum_{i=1}^m \text{Tr}\,(P_i u P_i T(x) P_i) \qquad \text{for all } x \in G,$$

where each $x \mapsto T(x)P_i$ is irreducible, so

$$\| \operatorname{Tr}(P_i u P_i T(\cdot) P_i) \|_A = \| P_i u P_i \|_1,$$

and

$$\| f \|_A \leq \sum_{i=1}^{m} \| P_i u P_i \|_1 \leq \| u \|_1 \text{ by 2.10. } \square$$

4.16 Theorem: *$A(G)$ is a Banach algebra under the pointwise operations and is closed under conjugation, and translation.*

Proof: We already know $A(G)$ is a Banach space closed under conjugation and right translation. As for left translation, put $L(x)f(y) = f(x^{-1}y)$ for $x, y \in G$, then $(L(x)f)\hat{}_\alpha = \hat{f}_\alpha T_\alpha(x^{-1})$ so

$$\| L(x)f \|_A = \| \hat{f}\hat{R}(x^{-1}) \|_1 = \| \hat{f} \|_1.$$

It remains to show that $\| fg \|_A \leq \| \mathrm{f} \|_A \| g \|_A$ for $f, g \in A(G)$. For any $\alpha, \beta \in \hat{G}$ we have

$$\begin{aligned} n_\alpha n_\beta \tilde{f}_\alpha(x) \tilde{g}_\beta(x) &= n_\alpha \operatorname{Tr}(T_\alpha(x)\hat{f}_\alpha) n_\beta \operatorname{Tr}(T_\beta(x)\hat{g}_\beta) \\ &= n_\alpha n_\beta \operatorname{Tr}((T_\alpha(x) \otimes T_\beta(x))(\bar{\hat{f}}_\alpha \otimes \hat{g}_\beta)), \end{aligned}$$

but $T_\alpha \otimes T_\beta$ is a finite dimensional unitary representation of G so 4.15 shows that

$$\| n_\alpha n_\beta \tilde{f}_\alpha \tilde{g}_\beta \|_A \leq n_\alpha n_\beta \| \hat{f}_\alpha \otimes \hat{g}_\beta \|_1 = n_\alpha n_\beta \| \hat{f}_\alpha \|_1 \| \hat{g}_\beta \|_1$$

(by 2.9). Thus

$$\begin{aligned} \| fg \|_A &\leq \sum_\alpha \sum_\beta \| n_\alpha n_\beta \tilde{f}_\alpha \tilde{g}_\beta \|_A \\ &\leq \sum_\alpha \sum_\beta n_\alpha n_\beta \| \hat{f}_\alpha \|_1 \| \hat{g}_\beta \|_1 \\ &= \| f \|_A \| g \|_A. \quad \square \end{aligned}$$

Observe that for characters $\| \chi_\alpha \|_A = \| \chi_\alpha \|_\infty = n_\alpha$, for $\alpha \in \hat{G}$, in particular $\| 1 \|_A = 1$.

4.17: The **dual space of A(G)** can be naturally identified with $\mathscr{L}^\infty(\hat{G})$ with the action $R_\phi(f) = \operatorname{Tr}(\phi \hat{f})$, $\phi \in \mathscr{L}^\infty$, $f \in A(G)$ (see 3.9). Also each $\phi \in \mathscr{L}^\infty$ defines a bounded operator on each of $A(G)$ and $L^2(G)$ by $\hat{\phi} f = \mathscr{F}^{-1}(\phi \mathscr{F} f)$. Then $R_\phi(f) = \hat{\phi} f(e)$ for $f \in A(G)$, and $\hat{\phi}\, L(x) = L(x)\hat{\phi}$, $x \in G$, where $(L(x)f)(y) = f(x^{-1}y)$ for all $y \in G$. Proposition 3.6 shows that the operator norm $\| \hat{\phi} \| = \| \phi \|_\infty$. Thus $\mathscr{L}^\infty$ can also be interpreted as an algebra of operators on each of $A(G)$ and $L^2(G)$, commuting with left translations.

5. The Maximal Ideal Space of A(G)

5.1: We will now show that the maximal ideal space of $A(G)$ can be naturally identified with G, that is, for each multiplicative linear functional R_ϕ on $A(G)$ there exists $x_\phi \in G$ such that $R_\phi f = f(x_\phi)$ for all $f \in A(G)$. The theory of regular function algebras and the support of a linear functional will be used in the proof.

5.2 Proposition: *Let K be compact $\subset G$, $K \subset U$ open, then there exists $f \in A(G)$ satisfying $f(x) = 1$ for $x \in K$, $0 \le f(x) \le 1$ for all $x \in G$, and* spt$f \subset U$ *(where* spt$f$ = *closure* $\{x \in G : f(x) \neq 0\}$).

Proof: Let V be a neighborhood of e such that $\overline{KV^{-1}V} \subset U$, and let $f = (m_G(V))^{-1}\chi_{KV^{-1}} * \chi_V$. Then $f(x) = m_G(KV^{-1} \cap xV^{-1})/m_G(V)$ for all $x \in G$, and $0 \le f \le 1$, $f = 1$ on K, $f(x) = 0$ for $x \notin KV^{-1}V$. Further $f \in A(G)$ by 4.13. □

This shows that $A(G)$ is a regular function algebra (see [L, p. 57]) and standard arguments (for example, [RRC, p. 40]) show that there exist partitions of unity in $A(G)$ which are subordinate to a given open covering of G.

5.3 Definition: Let $\phi \in \mathscr{L}^\infty(\hat{G})$, then the **support of** ϕ, denoted by spt ϕ, is defined to be $\cap\{K : K$ compact $\subset G$, $R_\phi(f) = 0$ whenever spt $f \subset K^c$, $f \in A(G)\}$. Equivalently, $x \in$ spt ϕ if and only if for each neighborhood V of x there exists $f \in A(G)$ such that spt $f \subset V$ and $R_\phi(f) \neq 0$. If $\phi = \hat{\mu}$ for some $\mu \in M(G)$, then $R_\phi(f) = \int_G f(x^{-1})\,d\mu(x)$, so spt $\hat{\mu} = (\text{spt}\,\mu)^{-1}$ (spt μ is the ordinary support of μ).

5.4 Proposition: *If $\phi \in \mathscr{L}^\infty$, $f \in A(G)$ and* spt $\phi \cap$ spt $f = \emptyset$ *then $R_\phi(f) = 0$. Hence if $\phi \neq 0$, $\phi \in \mathscr{L}^\infty$, then* spt $\phi \neq \emptyset$.

Proof: Each $x \in$ spt f has a neighborhood V_x such that $V_x \cap$ spt $\phi = \emptyset$ and spt $g \subset V_x$ implies $R_\phi(g) = 0$. By a partition of unity argument there exists a set $\{g_i\}_{i=1}^m$ in $A(G)$ such that $\sum_{i=1}^m g_i(x) = 1$ for all $x \in$ spt f, and spt $g_i \subset V_{x_i}$ for some $x_i \in$ spt f, each i. Thus

$$R_\phi(f) = R_\phi\left(\sum_{i=1}^m g_i f\right) = \sum_{i=1}^m R_\phi(g_i f) = 0,$$

since spt $g_i f \subset V_{x_i}$. □

5.5 Proposition: *If $\phi \in \mathscr{L}^\infty$ and R_ϕ is a (nonzero) multiplicative linear functional then* spt ϕ *is a single point.*

Proof: Suppose $x, y \in \text{spt}\ \phi$, $x \neq y$ then there exist disjoint open neighborhoods V_1, V_2 of x and y respectively, and functions f, g in $A(G)$ such that spt $f \subset V_1$, spt $g \subset V_2$ and $R_\phi(f) \neq 0$, $R_\phi(g) \neq 0$. But then $fg = 0$ and $R_\phi(fg) = R_\phi(f)R_\phi(g) \neq 0$, a contradiction. Thus by 5.4, spt ϕ is a single point. □

5.6 Lemma: *Let $\phi \in \mathscr{L}^\infty$ have the property that* spt $(\hat{\phi} f) \subset$ spt f *for each $f \in A(G)$ then $\phi = cI$, for some complex constant c.*

We will prove this in several steps. Observe that the hypothesis on ϕ is equivalent to the following: if E is open $\subset G$, $f = 0$ on E, $f \in A(G)$ then $\hat{\phi} f = 0$ on E.

5.6 (i): *If E open $\subset G, f \in A(G)$, and f is constant on E then $\hat{\phi} f$ is constant on E.*

Proof: Let $x, y \in E$ and let V be a neighborhood of e such that $xV \cup yV \subset E$. Let $V_1 = yV$. If $p \in V_1$ then $xy^{-1}p \in E$ and $p \in E$ so $f(p) = f(xy^{-1}p) = (L(yx^{-1})f)(p)$. Thus $f - L(yx^{-1})f = 0$ on V_1, hence $\hat{\phi} f - L(yx^{-1})\hat{\phi} f = 0$ on V_1 (see 4.17), that is, $\hat{\phi} f(p) = \hat{\phi} f(xy^{-1}p)$ for $p \in V_1$, in particular $\hat{\phi} f(y) = \hat{\phi} f(x)$.

5.6 (ii): *There exists a constant c depending only on ϕ such that for any open $E \subset G$, $\hat{\phi} f = c$ on E whenever $f \in A(G)$ and $f = 1$ on E.*

Proof: By 5.6 (i) $\hat{\phi} f(x) = c(f, E)$ for each $x \in E$, but $c(f, E)$ does not depend on f, for if $f = g = 1$ on E, then $f - g = 0$ on E so $\hat{\phi} f - \hat{\phi} g = 0$, that is, $\hat{\phi} f = \hat{\phi} g = c(E)$ on E. Now let E_1, E_2 be open $\subset G$, then

$$c(E_1) = c(1, E_1) = c(1, E_1 \cup E_2) = c(1, E_2) = c(E_2),$$

so c does not depend on E.

5.6 (iii): $\phi = cI$.

Proof: It suffices to show that $\hat{\phi}\chi_K = c\chi_K$ for any compact $K \subset G$, considering the action of $\hat{\phi}$ on $L^2(G)$, since $Sp\{\chi_K : K \text{ compact}\}$ is dense in $L^2(G)$. Let K be compact $\subset G$, and let $\{U_n\}$ be a sequence of open neighborhoods of K such that $m_G(U_n \backslash K) \to 0$ as $n \to \infty$ and $U_n \supset U_{n+1}$ for all n. For each n let V_n be an open set such that $K \subset V_n \subset \overline{V}_n \subset U_n$. Then by 5.2 let $g_n \in A(G)$ such that $0 \leq g_n \leq 1$, $g_n = 1$ on $\overline{V}_n$ and $g_n = 0$ off U_n. Thus $\| g_n - \chi_K \|_2 \leq [m_G(U_n \backslash K)]^{1/2}$, so $\| \hat{\phi} g_n - \hat{\phi}\chi_K \|_2 \to 0$ as $n \to \infty$. Further by (ii) $\hat{\phi} g_n = c$ on V_n and $\hat{\phi} g_n = 0$ off U_n. Thus $\hat{\phi} g_n(x) \xrightarrow{n} c\chi_K(x)$ for all $x \notin \cap_n(U_n \backslash K)$, namely, (m_G) a.e., so $\hat{\phi}\chi_K = c\chi_K$. □

5.7 Theorem: *Let R_ϕ be a nonzero multiplicative linear functional on $A(G)$, then $\phi = \hat{R}(x)$ for some $x \in G$, that is, $R_\phi(f) = f(x)$ for all $f \in A(G)$.*

Proof: By 5.5, spt $\phi = \{x\}$ for some $x \in G$. Let $\psi = \phi\hat{R}(x^{-1})$, then $R_\psi(fg) = R_\phi[R(x^{-1})(fg)] = R_\phi[(R(x^{-1})f)(R(x^{-1})g)] = R_\psi(f)R_\psi(g)$ so R_ψ is also multiplicative and spt $\psi = \{e\}$. Let $f \in A(G)$ and suppose $y \notin \text{spt} f$, then $e \notin \text{spt}\, L(y^{-1})f$, thus $\hat{\psi} f(y) = R_\psi(L(y^{-1})f) = 0$ and spt $\hat{\psi} f \subset \text{spt} f$. Hence by Lemma 5.6, $\psi = cI = c\hat{R}(e)$ for some complex c, so $R_\psi(f) = cf(e)$ for all $f \in A(G)$, but R_ψ is multiplicative, so $c = 1$. Finally $\phi = \psi\hat{R}(x) = \hat{R}(x)$. □

Thus the maximal ideal space of $A(G)$ is identified with G in the above way, the spectral norm of $f \in A(G)$ is the sup-norm $\| f \|_\infty$, and if $f(x) \neq 0$ for all $x \in G$, then $1/f \in A(G)$.

6. Functions that Operate in A(G)

6.1: Suppose E is closed in **C**, and ψ is a complex function on E, then we say ψ **operates in A(G)** if $f \in A(G)$, range $f \subset E$ implies $\psi \circ f \in A(G)$. The Arens-Calderón-Shilov theorem [Ri, p. 161] shows that if E is convex then real-analytic functions operate. If G has an infinite abelian subgroup, for example, if G is a Lie group, then this condition is also necessary; that is, if E is convex and ψ operates in $A(G)$, then ψ is real analytic on a neighborhood of E. This will be proved by lifting the known result to G from an infinite compact abelian subgroup H. In the sequel H will be a closed subgroup of G, m_H its normalized Haar measure, and ρ the restriction map: $C(G) \to C(H)$.

6.2 Proposition: *Let $f \in A(G)$ then $\rho f \in A(H)$ and $\| \rho f \|_A \leq \| f \|_A$.*

Proof: Now $f(x) = \Sigma_\alpha n_\alpha \operatorname{Tr}(T_\alpha(x)\hat{f}_\alpha)$ for all $x \in G$, so $\rho f(h) = \Sigma_\alpha n_\alpha \operatorname{Tr}(T_\alpha (h)\hat{f}_\alpha)$. But $T_\alpha | H$ is a finite-dimensional unitary continuous representation of H, so by 4.15, $\| \operatorname{Tr}(T_\alpha(h)\hat{f}_\alpha) \|_A \leq \| \hat{f}_\alpha \|_1$. Thus the series $\Sigma_\alpha n_\alpha \operatorname{Tr}(T_\alpha(h)\hat{f}_\alpha)$ converges absolutely in $A(H)$, and $\| \rho f \|_A \leq \| f \|_A$. □

6.3: To show that each $f \in A(H)$ is a restriction of some function in $A(G)$ it is necessary to consider **induced representations.** Let $\hat{H}$ be the dual of H (see 7.2.5), and let τ_i be an element of the class $i \in \hat{H}$, with character ϕ_i and dimension n_i. For $\alpha \in \hat{G}$, $T_\alpha | H$ is a continuous unitary representation of H, hence it is a direct sum of irreducible representations of H, say $T_\alpha | H \cong \Sigma \oplus_{i \in \hat{H}} N_\alpha(i)\, \tau_i$, where $N_\alpha(i)$ is a natural number and $N_\alpha(i) > 0$ for only a finite number of $i \in \hat{H}$, since $n_\alpha = \Sigma_i N_\alpha(i) n_i$.

For $i \in \hat{H}$, let $\phi_i' \in M(G)$ be defined by $\int_G f\, d\phi_i' = \int_H f \phi_i\, dm_H$ for all $f \in C(G)$, then ϕ_i' is called the **generalized character** of the representation of

G which is induced by τ_i. Now for $\alpha \in \hat{G}$,

$$(\phi_i')\hat{}_\alpha = \int_H T_\alpha(h^{-1})\phi_i(h)\, dm_H(h),$$

thus $(\phi_i')\hat{}_\alpha \neq 0$ if and only if $N_\alpha(i) > 0$, but $\phi_i' \neq 0$ in $M(G)$ therefore there exists at least one $\alpha \in \hat{G}$ such that $N_\alpha(i) > 0$. Now choose one such $\alpha \in \hat{G}$ for each $i \in \hat{H}$, and denote it by $\alpha(i)$, but also subject to the requirement that $\alpha(\bar{i}) = \overline{\alpha(i)}$. This latter is possible, since $T_{\bar{\alpha}} | H \cong \Sigma \oplus_i N_\alpha(i)\tau_{\bar{i}}(h)$.

6.4 Theorem: *Let $f \in A(H)$ then there exists $F \in A(G)$ such that $\rho F = f$ and $\| F \|_A = \| f \|_A$. Further if f is real, then F may be chosen to be real also.*

Proof: If $f \in A(H)$ then $f(h) = \Sigma_i n_i \operatorname{Tr}(\tau_i(h)\hat{f}_i)$, for all $h \in H$, thus it suffices to consider f of the form $h \mapsto n_i \operatorname{Tr}(\tau_i(h)\hat{f}_i)$, some $i \in \hat{H}$. Let $\alpha = \alpha(i)$ then we may choose a basis for the space X_α (on which T_α acts) such that

$$T_\alpha(h) = \Sigma \oplus_j N_\alpha(j)\tau_j(h).$$

Let $\hat{F}_\alpha \in \mathscr{B}(X_\alpha)$ be defined by $(n_i/n_\alpha)(\hat{f}_i \oplus 0)$, where the direct sum is taken so that $n_\alpha \operatorname{Tr}(T_\alpha(h)\hat{F}_\alpha) = n_i \operatorname{Tr}(\tau_i(h)\hat{f}_i)$ for all $h \in H$. Then let

$$F_i(x) = n_\alpha \operatorname{Tr}(T_\alpha(x)\hat{F}_\alpha),$$

all $x \in G$, thus $\rho F_i = f$, and

$$\| F_i \|_A = n_\alpha \| \hat{F}_\alpha \|_1 = n_i \| \hat{f}_i \|_1 = \| f \|_A.$$

For any $f \in A(H)$, there exists for each $i \in \hat{H}$ a function $F_i \in A(G)$ such that $F_i(h) = n_i \operatorname{Tr}(\tau_i(h)\hat{f}_i)$ for all $h \in H$, and $\| F_i \|_A = n_i \| \hat{f}_i \|_1$. Let $F = \Sigma_i F_i$, which is absolutely convergent in $A(G)$ since

$$\| F \|_A \leq \sum_i \| F_i \|_A = \sum_i n_i \| \hat{f}_i \|_1 = \| f \|_A.$$

But $\rho F = f$, so $\| f \|_A \leq \| F \|_A$ (by 6.2), thus $\| F \|_A = \| f \|_A$. Further if f is real, then the direct sums $(\hat{f}_i \oplus 0)$ may be taken so that $J(\mathscr{F} F)J = \mathscr{F} F$, (see 4.2) that is, $J\hat{F}_\alpha J = \hat{F}_{\bar{\alpha}}$, for each $\alpha \in \hat{G}$, since f satisfies these relations (for the map J defined on H) and $\alpha(\bar{i}) = \overline{\alpha(i)}$, thus F is real. □

6.5 Theorem: *Suppose G has an infinite abelian subgroup then only real-analytic functions operate in $A(G)$.*

Proof: Let H be a compact infinite abelian subgroup of G. Then for $r > 0$

$$\sup \{ \| e^{if} \|_A : \text{real } f \in A(H), \| f \|_A \leq r \} = e^r,$$

by [R, p. 143, 6.6.2]. For any $\varepsilon > 0$ there exists real $f \in A(H)$, such that $\| f \|_A \leq r$ and $\| e^{if} \|_A > e^r - \varepsilon$. Then by Theorem 6.4 there exists real $F \in A(G)$ such that $\rho F = f$ and $\| F \|_A = \| f \|_A \leq r$. But $\rho e^{iF} = e^{if}$ so $e^r \geq \| e^{iF} \|_A \geq \| e^{if} \|_A > e^r - \varepsilon$, thus

$$\sup \{ \| e^{iF} \|_A : \text{real } F \in A(G), \| F \|_A \leq r \} = e^r.$$

Now the proof of Helson, Kahane, Katznelson, and Rudin [1] applies to $A(G)$ to show that only real-analytic functions operate (this proof can also be found in [R, p. 144–6, 6.6.3; p. 149, 6.9.3]). □

7. Remarks

7.1: A W^*-algebra is a $*$-algebra of operators on a Hilbert space which contains the identity and is closed in the strong operator topology. It can be shown that $\mathscr{L}^\infty(\hat{G})$ is the W^*-algebra generated by $\{R(x): x \in G\}$ acting on $L^2(G)$, or $\{\hat{R}(x): x \in G\}$ on $\mathscr{L}^2(\hat{G})$. We defined a trace on $\mathscr{L}^\infty(\hat{G})$ (actually on a subspace of $\mathscr{L}^\infty$), and $\mathscr{L}^1(\hat{G})$ is the trace class, $\mathscr{L}^2(\hat{G})$ the class of Hilbert–Schmidt operators with respect to this trace. As in the standard theorems on traces, $\mathscr{L}^1$ is the dual of the space of compact operators $\mathscr{C}_0$, and $\mathscr{L}^\infty$ is the dual of the trace class $\mathscr{L}^1$. These concepts may be extended to some locally compact, noncompact groups without too much detailed knowledge of the dual of the group (which is still the set of equivalence classes of all continuous unitary irreducible, but not necessarily finite-dimensional, representations). This comes about because, as stated above, $\mathscr{L}^\infty$ may be defined with reference to the quite accessible space $L^2(G)$, rather than to the dual which in general (that is, non-compact) is rather inaccessible.

In Appendix B we give a concise presentation of the theory of $\mathscr{L}^p(\hat{G})$, $1 < p < \infty$. Hölder and Minkowski-type inequalities are valid in this setting as well as a version of the Hausdorff–Young theorem.

7.2 Historical Notes: Krein [1, 2] first studied $A(G)$ from the point of view of positive definite functions and found its maximal ideal space. (Observe that if $f \in A(G)$ then a standard identity shows that f is a linear combination of four positive definite functions of the form $g * g^*$, $g \in L^2(G)$).

Eymard [1] studied the W^*-algebra $VN(G)$ (our $\mathscr{L}^\infty$) generated by $\{L(x): x \in G\}$ on $L^2(G, dm_l)$ for locally compact G, left-invariant Haar measure m_l, and interpreted $A(G)$ as the pre-dual of $VN(G)$. He showed that the maximal ideal space of $A(G)$ can be identified with G, and it is his proof, restricted to compact groups, which is given in 5.5, 5.6, and 5.7.

The trace idea was pursued for unimodular locally compact groups (Haar measure is left and right invariant) by Segal [2, 3] and Stinespring [1]. Stinespring showed that any symmetric (that is, $\phi(\bar{f}) = \overline{\phi(f)}$) multiplicative linear functional on $A(G)$ is a point-evaluation on G. Another W^*-algebra type of proof of the maximal ideal theorem can be found in Saitô [1], again, however, subject to the assumption of symmetry.

Section 6 is from Dunkl [3]. As we mention in Appendix C, Rider [5] has recently proved Theorem 6.5 for any infinite compact group without the (possibly vacuous) restriction that there is an infinite abelian subgroup.

CHAPTER 9

HOMOGENEOUS SPACES

1. Introduction

1.1: The machinery developed in Chapter 7 can also be applied to compact homogeneous spaces, which are essentially spaces of right cosets of a closed, not necessarily normal, subgroup of a compact group G. The theory includes for example spherical harmonics, the Poisson integral for the sphere, orthogonal polynomials, and analytic functions on the ball.

Sections 2 through 5 are a straightforward application of the Fourier series principles and expound on L^p-theory on homogeneous spaces, convolution operators, and spherical functions, which play the part of the characters.

In Section 6 we have the somewhat unexpected result that the set of measures on $\mathbf{SO}(n)$ $(n \geq 3)$, bi-invariant under $\mathbf{SO}(n-1)$, forms a commutative algebra with an uncomplicated maximal ideal space and few idempotents. We show also how a certain family of orthogonal polynomials, namely Gegenbauer or ultraspherical polynomials, appears in a natural way on $\mathbf{SO}(n)$.

Throughout this chapter, **G is a compact group, and H is a closed subgroup,** with normalized Haar measure m_H, which will often be viewed as an element of $M(G)$.

2. Basic Definitions

2.1 Definition: $G/H = \{Hx : x \in G\}$, the space of right cosets with the quotient topology. Let ρ be the map $G \to G/H$ given by $\rho x = Hx$. There is a continuous map of $G/H \times G \to G/H$ given by $(Hx, y) \mapsto Hxy$; note that any coset may be moved to any other coset this way, $Hx(x^{-1}y) = Hy$.

2.2 Definition: A **homogeneous space** for G is a compact space X for which there is a continuous mapping $X \times G \to X$ denoted by $(s, x) \mapsto sx$, such that $se = s$, $(sx)y = s(xy)$, for all $s \in X$, $x, y \in G$ and for any $s, t \in X$ there exists $x \in G$ such that $sx = t$.

2.3 Proposition: *Let X be a homogeneous space for G and let $p \in X$, then X is isomorphic to G/H where $H = \{x \in G: px = p\}$.*

Proof: Let H be as stated then H is a closed subgroup of G and G/H is defined as in 2.1. Define the map $\rho_1: G \to X$ by $\rho_1 x = px$. Then $\rho_1(xy) = p(xy) = (\rho_1 x)y$ and ρ_1 is onto X. For any $s \in X$, $\rho_1^{-1} s$ is a right coset of H, since $px = py = s$ implies

$$pxy^{-1} = p, \ xy^{-1} \in H, \quad x \in Hy,$$

and in fact ρ_1^{-1} is a one-to-one map of X onto G/H, and is the required isomorphism. □

Thus it suffices to study the spaces G/H.

2.4 Definition: Let

$$C_H(G) = \{f \in C(G): f(hx) = f(x) \text{ for all } x \in G, h \in H\}.$$

Then $C_H(G)$ is a closed subspace of $C(G)$; and the condition $f(hx) = f(x)$ is equivalent to $m_H * f = f$, that is, $f(x) = \int_H f(hx)\, dm_H(h)$, $(x \in G)$. Thus for $1 \leq p \leq \infty$ we define

$$L_H^p(G) = \{f \in L^p(G): m_H * f = f\}$$

$$M_H(G) = \{\mu \in M(G): m_H * \mu = \mu\}.$$

Each of the spaces $C_H(G)$, $L_H^p(G)$, $M_H(G)$ is a closed subspace invariant under R (right translation) and is the range of a bounded projection of the appropriate space, namely $\pi(f) = m_H * f$, since $m_H * m_H = m_H$.

2.5 Proposition: *There is a strongly continuous representation of G defined on $C(G/H)$ by $R(x)f(Hy) = f(Hyx)$ $(x, y \in G)$ and $(R, C(G/H)) \cong (R, C_H(G))$.*

Proof: Let $f \in C(G/H)$, then $f \circ \rho \in C_H(G)$ since $f \circ \rho\ (x) = f(Hx)$, and $f \circ \rho\ (hx) = f(Hhx) = f(Hx)$ $(x \in G, h \in H)$. The map $f \mapsto f \circ \rho$ is one-to-one, and if $g \in C_H(G)$ then the function $f\colon Hx \mapsto g(x)$ is well defined and continuous on G/H with $g = f \circ \rho$. Finally $(R(x)f) \circ \rho\ (y) = R(x)f(Hy) = f(Hyx) = R(x)f(Hy) = R(x)\,(f \circ \rho)\,(y)$. □

It will generally be more convenient to work with $C_H(G)$ than with $C(G/H)$.

2.6 Proposition: *The dual space of $C_H(G)$ is naturally identified with $M_H(G)$, so there is a natural correspondence between $M_H(G)$ and $M(G/H)$.*

Proof: Let $\mu \in M_H(G)$ and let $S\mu$ be the linear functional on $C_H(G)$ given by $S\mu(f) = \int_G f\, d\mu$, then $\| S\mu \| \leq \| \mu \|$. Further suppose that $S\mu = 0$ and let $f \in C(G)$, then

$$\int_G f\, d\mu = \int_G f\, d(m_H * \mu) = \int_G \int_H f(hx)\, dm_H(h)\, d\mu(x)$$
$$= \int_G \pi(f)\, d\mu = 0,$$

since $\pi(f) \in C_H(G)$, thus $\mu = 0$.

Conversely, given a bounded linear functional λ on $C_H(G)$, then by the Hahn–Banach and Riesz theorems there exists $\mu_1 \in M(G)$ such that $\lambda(f) = \int_G f\, d\mu_1$ for all $f \in C_H(G)$ and $\| \mu_1 \| = \| \lambda \|$. But then $\lambda(f) = \int_G f(hx)\, d\mu_1(x)$ for all $h \in H$, so

$$\lambda(f) = \int_G \int_H f(hx)\, dm_H(h)\, d\mu_1(x) = \int_G f\, d(m_H * \mu_1).$$

Now let $\mu = m_H * \mu_1$, then $\mu \in M_H(G)$, $S\mu = \lambda$, and

$$\| \mu \| \leq \| \mu_1 \| = \| \lambda \| \leq \| \mu \|. \quad \square$$

By a similar argument the dual of $L^p_H(G)$ is $L^q_H(G)$ for $1 \leq p < \infty$, $1/p + 1/q = 1$.

2.7 Proposition: *There exists a unique normalized R-invariant $\omega \in M(G/H)$ and for $f \in C(G/H)$, $\int_{G/H} f\, d\omega = \int_G f(Hx)\, dx$.*

Proof: Let $\mu \in M(G/H)$ and be normalized and R-invariant, that is, $\mu(G/H) = 1$ and $\int_{G/H} R(x) f\, d\mu = \int_{G/H} f\, d\mu$, for all $x \in G$. By 2.6, $\int_{G/H} f\, d\mu = \int_G f(Hx)\, d\mu_1(x)$ for some $\mu_1 \in M_H(G)$, all $f \in C(G/H)$. Then the bounded linear functional $f \mapsto \int_G f\, d\mu_1$ on $C(G)$ is R-invariant and normalized, since $\int_G f\, d\mu_1 = \int_G \pi(f)\, d\mu_1$ and μ_1 is R-invariant. Hence by the uniqueness of Haar measure $\mu_1 = m_G$. $\square$

A consequence of this is the formula

$$\int_G f\, dm_G = \int_{G/H} \pi(f)(x)\, d\omega(Hx), \text{ for all } f \in C(G).$$

2.8 Definition: For $1 \leq p \leq \infty$, let $L^p(G/H) = L^p(G/H, d\omega)$. Then clearly $L^p(G/H) \cong L^p_H(G)$.

3. Operators that Commute with Translations

3.1: Let (T, X) be a representation of G on a Banach space X, then we say S is **G-operator** on X, if $S \in \mathscr{B}(X)$ and $ST(x) = T(x)S$ for all $x \in G$. In the following section we will study G-operators on $C_H(G)$, $L^1_H(G)$ and $L^\infty_H(G)$. Observe for $\mu \in M(G)$, $f \in L^1(G)$ that $\mu * R(x)f = R(x)(\mu * f)$.

3.2 Theorem: *Let S be a G-operator on $C_H(G)$ then there exists a unique $\mu \in M_H(G)$ such that $\mu * m_H = \mu$, $Sf = \mu * f$ for all $f \in C_H(G)$ and $\| \mu \| = \| S \|$.*

Proof: The map $f \mapsto Sf(e)$ is a bounded linear functional on $C_H(G)$, hence by 2.6 there exists a unique $\mu_1 \in M_H(G)$ such that $Sf(e) = \int_G f \, d\mu_1$ and $\| \mu_1 \| \leq \| S \|$, since $| Sf(e) | \leq \| S \| \; \| f \|_\infty$. Further, S is a G-operator so $Sf(x) = (R(x)Sf) \; (e) = S(R(x)f) \; (e) = \int_G f(yx) \; d\mu_1(y) = \int_G f(y^{-1}x) \; d\mu(y)$ for all $x \in G$, where $\mu(E) = \mu_1(E^{-1})$, for E Borel $\subset G$. Thus $Sf = \mu * f$, and $\mu * m_H = \mu$ since $m_H * \mu_1 = \mu_1(\mu_1 \in M_H(G))$. But $S\pi(f) \in C_H(G)$ for all $f \in C(G)$, so $m_H * [\mu * (m_H * f)] = (\mu * m_H) * f = \mu * f$, therefore $m_H * \mu * m_H = \mu$. Further, $\| Sf \| \leq \| \mu \| \; \| f \|$, hence $\| S \| \leq \| \mu \| \leq \| S \|$. □

3.3 Definition: Let $M_{HH}(G) = \{\mu \in M(G) : m_H * \mu * m_H = \mu\}$. Then $M_{HH}(G)$ is a closed subspace of $M_H(G)$, in general not R-invariant. In fact $M_{HH}(G)$ is a Banach $*$-algebra with the identity m_H and is closed under the operation $\mu \mapsto \check{\mu}$, $\check{\mu}(E) = \mu(E^{-1})$, E Borel $\subset G$.

$C_{HH}(G)$ and $L^p_{HH}(G)$ are similarly defined. Elements of these spaces are said to be bi-invariant under H.

3.4 Remark: The various convolution inequalities (see 7.1.4) can be applied to products of the sort $g * f$, where $m_H * g * m_H = g$, $m_H * f = f$, and $m_H * (g * f) = g * f$.

3.5 Lemma: *Suppose $f \in L^\infty(G)$ and $x \mapsto R(x)f$ is strongly continuous, $G \to L^\infty(G)$, then $f \in C(G)$, that is, $f = g$ a.e. where $g \in C(G)$.*

Proof: Let $X = \overline{Sp}\{R(x)f : x \in G\}$, then (R, X) is a strongly continuous representation of G, hence by 7.4.5 is completely decomposable. That is, f is a $\| \cdot \|_\infty$-limit of trigonometric polynomials, hence continuous. □

3.6 Theorem: *Let S be a G-operator on $L^1_H(G)$ then there exists a unique $\mu \in M_{HH}(G)$ such that $Sf = \mu * f$ for all $f \in L^1_H(G)$.*

Proof: Now S^* is a G-operator on $L^\infty_H(G)$ for which $C_H(G)$ is an invariant subspace. For if $f \in C_H(G)$ then

$$\| R(x)S^*f - S^*f \|_\infty = \| S^*R(x)f - S^*f \|_\infty \leq \| S^* \| \; \| R(x)f - f \|_\infty \to 0$$

as $x \to e$, so by 3.5, $S^*f \in C_H(G)$. Hence by 3.2 there exists a unique $\nu \in M_{HH}(G)$ such that $S^*f = \nu * f$ for all $f \in C_H(G)$. For $g \in L^1_H(G)$, $f \in C_H(G)$ we have $\int_G g(\nu * f) \, dm_G = \int_G (\check{\nu} * g) f \, dm_G$, so we put $\mu = \check{\nu}$, then $\mu \in M_{HH}(G)$ and $Sg = \mu * g$. □

3.7 Multiplier transformations: Suppose $\phi = \{\phi_\alpha\colon \phi_\alpha \in \mathscr{B}(X_\alpha),\ \alpha \in \hat{G}\}$ has the property that $\hat{f}\phi \in \mathscr{F}C_H(G)$ whenever $f \in C_H(G)$, then the map S: $C_H(G) \to C_H(G)$, where $(Sf)\hat{} = \hat{f}\phi$, is a G-operator, and hence $\hat{f}_\alpha \phi_\alpha = \hat{f}_\alpha \hat{\mu}_\alpha$ all $\alpha \in \hat{G}$, some $\mu \in M_{HH}(G)$ (this implies $(m_H)\hat{}_\alpha \phi_\alpha = \hat{\mu}_\alpha$). The map S is closed, for if $f_n \xrightarrow{n} f$, $Sf_n \xrightarrow{n} g$ in $C_H(G)$, then for each $\alpha \in \hat{G}$ $(\mathscr{F}f_n)_\alpha \xrightarrow{n} (\mathscr{F}f)_\alpha$ and $(\mathscr{F}Sf_n)_\alpha = (\mathscr{F}f_n)_\alpha \phi_\alpha \xrightarrow{n} (\mathscr{F}g)_\alpha$ so $(\mathscr{F}f)_\alpha \phi_\alpha = (\mathscr{F}g)_\alpha$, thus $g = Sf$. By the closed graph theorem S is bounded, and since $\mathscr{F}(R(x)f)_\alpha = T_\alpha(x)\hat{f}_\alpha$, S is a G-operator.

A similar theorem holds for multiplier transformations on $L^1_H(G)$ and $L^\infty_H(G)$. For $L^\infty_H(G)$ we use Lemma 3.5 and the fact that a multiplier transformation on $L^\infty_H(G)$ is determined by its action on $C_H(G)$.

4. Spherical Functions

4.1: If $f \in L^1_H(G)$, $\mu \in M_H(G)$ then for $\alpha \in \hat{G}$

$$\tilde{f}_\alpha = \chi_\alpha * f = \chi_\alpha * (m_H * f) = (\chi_\alpha * m_H) * f,$$

and similarly

$$\tilde{\mu}_\alpha = (\chi_\alpha * m_H) * \mu.$$

Further, $\chi_\alpha * m_H = m_H * \chi_\alpha$ (see 7.3.6), so $m_H * \tilde{f}_\alpha = \tilde{f}_\alpha$ and $m_H * \tilde{\mu}_\alpha = \tilde{\mu}_\alpha$ implying $\tilde{f}_\alpha$ and $\tilde{\mu}_\alpha \in C_H(G)$.

We thus define the **spherical function** $\phi_\alpha = \chi_\alpha * m_H$, for $\alpha \in \hat{G}$. The Fourier series of m_H is $\Sigma_\alpha n_\alpha \phi_\alpha$.

4.2 Proposition: *For $\alpha \in \hat{G}$, ϕ_α has the following properties:*

(i) $\phi_\alpha \in C_{HH}(G)$
(ii) $\phi_\alpha * \phi_\beta = \delta_{\alpha\beta}\phi_\alpha/n_\alpha$
(iii) $\phi_\alpha(x^{-1}) = \overline{\phi_\alpha(x)}$ *for all* $x \in G$, *and* $\phi_{\bar{\alpha}} = \overline{\phi_\alpha}$
(iv) $\phi_\alpha(e) = m_\alpha$, *a nonnegative integer,*

$$\| \phi_\alpha \|_2 = (m_\alpha/n_\alpha)^{1/2}, \quad \textit{and } \| \phi_\alpha \|_\infty = m_\alpha.$$

Proof: The first three are easily derived from 7.3.6. Now $\phi_\alpha(e) = \int_H \chi_\alpha(h)dm_H(h)$ so by 7.3.8, $\phi_\alpha(e)$ is the number of times that $T_\alpha \mid H$ contains the trivial representation $H \to \{1\}$, which is a non negative integer; this number is the dimension of $\{\xi \in X_\alpha\colon T_\alpha(h)\,\xi = \xi$, for all $h \in H\}$. Further,

$$\phi_\alpha * \phi_\alpha(e) = \int_G \phi_\alpha(x)\phi_\alpha(x^{-1})\,dx = \int_G |\phi_\alpha(x)|^2\,dx = \phi_\alpha(e)/n_\alpha = m_\alpha/n_\alpha,$$

and

$$|\phi_\alpha(x)| = n_\alpha \left| \int_G \phi_\alpha(xy)\overline{\phi_\alpha(y)}\,dy \right| \le n_\alpha \| \phi_\alpha \|_2^2 = m_\alpha. \quad \square$$

Observe that (iv) shows that $\phi_\alpha = 0$ if $m_\alpha = 0$, or to put it another way, $\phi_\alpha \neq 0$ if and only if X_α contains at least one $\xi \neq 0$ such that $T_\alpha(h)\,\xi = \xi$ for all $h \in H$ (ξ is an eigenvector of eigenvalue 1 common to all the $T_\alpha(h)$'s).

4.3: Suppose now that H has the property that $m_\alpha = 0$ or 1 for any $\alpha \in \hat{G}$, then it turns out that $M_{HH}(G)$ is a commutative algebra. Observe that H does not necessarily have this property, for example, $H = \{e\}$ and $n_\alpha > 1$, but interesting examples exist and will be studied later, namely $G = \mathbf{SO}(n)$, $H = \mathbf{SO}(n-1)$ and $G = \mathbf{U}(n)$, $H = \mathbf{U}(n-1)$.

4.4 Lemma: *Let X be an R-invariant closed subspace of $C_H(G)$ such that $\dim (X \cap C_{HH}(G)) = 1$, then $X = n_\alpha\, \phi_\alpha * X$ for some $\alpha \in \hat{G}$ and $m_\alpha = 1$, and further, X is irreducible.*

Proof: Let X_1 be an irreducible subspace of X, then (R, X_1) is equivalent to some (T_α, X_α), $\alpha \in \hat{G}$. Since $X_1 \subset C_H(G)$ we must have $\phi_\alpha \neq 0$, so $m_\alpha \neq 0$. But by hypothesis $\dim (X_1 \cap C_{HH}(G)) \leq 1$ so $m_\alpha = 1$ (since $f \in C_{HH}(G)$ implies $R(h)f = f$ for all $h \in H$). If $X_1 \neq X$, then the orthogonal complement of X_1 in X has an irreducible component which must contain a nonzero function fixed under all $R(h)$, $h \in H$, contrary to the dimension hypothesis. Any $f \in X$ has the Fourier series $\Sigma_\beta\, n_\beta\, \phi_\beta * f = n_\alpha\, \phi_\alpha * f$, since $(R, X) \cong (T_\alpha, X_\alpha)$. □

4.5: Suppose now that $\alpha \in \hat{G}$ and $m_\alpha = 1$, then we can choose an orthonormal basis $\{\xi_j\}_{j=1}^{n_\alpha}$ for X_α such that $T_\alpha(h)\xi_1 = \xi_1$ for all $h \in H$ and $\int_H T_\alpha(h)\,\xi_j \, dm_H(h) = 0$ for $j > 1$, that is,

$$T_\alpha(h)_{11} = 1,\ T_\alpha(h)_{1j} = T_\alpha(h)_{j1} = 0$$

for $j > 1$, for all $h \in H$ and

$$\int_H T_\alpha(h)_{ij}\, dm_H(h) = 0$$

for $2 \leq i, j \leq n_\alpha$. This means that $\hat{(m_H)}_{\alpha ij} = \delta_{i1}\delta_{j1}$, $1 \leq i, j \leq n_\alpha$. Further, $\phi_\alpha = T_{\alpha 11}$. Now let $f \in L^1_H(G)$, then

$$\tilde{f}_\alpha(x) = \phi_\alpha * f(x) = \int_G T_\alpha(xy^{-1})_{11} f(y)\, dy = \sum_{j=1}^{n_\alpha} \hat{f}_{\alpha j1}\, T_\alpha(x)_{1j}$$

(see 7.3.7). We saw in 7.3.1 that $Sp\{T_{\alpha 1j}\colon 1 \leq j \leq n_\alpha\}$ is an irreducible subspace of $C(G)$, and the above indicates it is the unique subspace of $C_H(G)$ which is equivalent to (T_α, X_α). Applying this to the situation of 4.4, we see that $X = Sp\{T_{\alpha 1j}\colon 1 \leq j \leq n_\alpha\}$ and $T_{\alpha 11} = \phi_\alpha$ is the unique element of $X \cap C_{HH}(G)$, normalized by $\phi_\alpha(e) = 1$.

4.6 Proposition: *Let $\alpha \in \hat{G}$ with $m_\alpha = 1$ then there exists a multiplicative linear functional c_α on $M_{HH}(G)$ such that $\tilde{\mu}_\alpha = c_\alpha(\mu)\phi_\alpha$, $\| \hat{\mu}_\alpha \|_\infty = | c_\alpha(\mu) |$, $c_\alpha(\mu^*) = \overline{c_\alpha(\mu)}$, and $c_\alpha(\mu) = \int_G \overline{\phi_\alpha}\, d\mu$ for each $\mu \in M_{HH}(G)$.*

Proof: By 4.5, $\tilde{\mu}_\alpha(x) = \Sigma_{j=1}^{n_\alpha} \hat{\mu}_{\alpha j1} T_{\alpha 1 j}$, but $\mu * m_H = \mu$, so

$$\tilde{\mu}_\alpha = \phi_\alpha * \mu * m_H = \tilde{\mu}_\alpha * m_H,$$

hence

$$\tilde{\mu}_\alpha(x) = \sum_{j=1}^{n_\alpha} \hat{\mu}_{\alpha j1} \sum_{k=1}^{n_\alpha} T_\alpha(x)_{1k} \int_H T_\alpha(h)_{kj}\, dm_H(h) = \hat{\mu}_{\alpha 11} T_\alpha(x)_{11} = c_\alpha(\mu)\, \phi_\alpha(x),$$

$$c_\alpha(\mu) = \hat{\mu}_{\alpha 11} = \int_G \overline{T_\alpha(x)_{11}}\, d\mu(x).$$

Further, if $\mu, \nu \in M_{HH}(G)$, then

$$(\mu * \nu)\tilde{}_\alpha = \phi_\alpha * \mu * \nu = \tilde{\mu}_\alpha * \nu = c_\alpha(\mu)\,(\phi_\alpha * \nu) = c_\alpha(\mu) c_\alpha(\nu) \phi_\alpha,$$

but $\mu * \nu \in M_{HH}(G)$ so $(\mu * \nu)\tilde{}_\alpha = c_\alpha(\mu * \nu)\phi_\alpha$; thus c_α is multiplicative. Also

$$c_\alpha(\mu^*) = \int_G \overline{\phi_\alpha(x)}\, d\mu^*(x) = \int_G \overline{\phi_\alpha(x^{-1})\, d\mu(x)} = \overline{\int_G \overline{\phi_\alpha(x)}\, d\mu(x)} = \overline{c_\alpha(\mu)}. \quad \square$$

4.7 Proposition: *Let $\alpha \in \hat{G}$, $m_\alpha = 1$, $f \in L_H^1(G)$, $\mu \in M_{HH}(G)$, then $(\mu * f)\tilde{}_\alpha = c_\alpha(\mu) \tilde{f}_\alpha$.*

4.8 Proposition: *Suppose for each $\alpha \in \hat{G}$, $m_\alpha = 0$ or 1, then $M_{HH}(G)$ is a commutative Banach algebra and $L_{HH}^1(G)$ is a closed ideal in it.*

Proof: Let $\hat{G}_0 = \{\alpha \in \hat{G} : m_\alpha = 1\}$. Then $\mu \in M_{HH}(G)$ has the Fourier series $\Sigma_{\alpha \in \hat{G}_0} n_\alpha c_\alpha(\mu) \phi_\alpha$, and so if $\nu \in M_{HH}(G)$ then $\mu * \nu$ has the series $\Sigma_{\alpha \in \hat{G}_0} n_\alpha c_\alpha(\mu) c_\alpha(\nu) \phi_\alpha$, thus $\mu * \nu = \nu * \mu$. Finally $L_{HH}^1(G) = M_{HH}(G) \cap L^1(G)$ (identifying $f \in L^1(G)$ with $f\, dm_G$ in $M(G)$). $\square$

4.9 Theorem: *Suppose for each $\alpha \in \hat{G}$, $m_\alpha = 0$ or 1, then the maximal ideal space of $L_{HH}^1(G)$ is identified with $\hat{G}_0$, that is, each nonzero multiplicative linear functional is c_α for some $\alpha \in \hat{G}_0$.*

Proof: Let λ be a nonzero multiplicative linear functional on $L_{HH}^1(G)$, then by 2.6 there exists $g \in L_{HH}^\infty(G)$ such that $\lambda(f) = \int_G f\, \bar{g}\, dm_G$ for all $f \in L_{HH}^1(G)$. Now $\phi_\alpha * \phi_\beta = \delta_{\alpha\beta} \phi_\alpha / n_\alpha$ so $\lambda(\phi_\alpha * \phi_\beta) = \lambda(\phi_\alpha)\lambda(\phi_\beta) = \delta_{\alpha\beta}\lambda(\phi_\alpha)/n_\alpha$, for all $\alpha, \beta \in \hat{G}_0$. If $\lambda(\phi_\alpha) = 0$ for all α then $\int_G \phi_\alpha \bar{g}\, dm_G = 0$ so by 4.6, $g = 0$. Thus there exists $\alpha \in \hat{G}_0$ such that $\lambda(\phi_\alpha) \neq 0$; then the above formula shows $\lambda(\phi_\alpha) = 1/n_\alpha$ and $\lambda(\phi_\beta) = 0$ for $\beta \neq \alpha$, thus $g = \phi_\alpha$ and $\lambda(f) = \int_G f \overline{\phi_\alpha}\, dm_G = c_\alpha(f)$. $\square$

We already know that $\hat{f} \in \mathscr{C}_0(\hat{G})$ for $f \in L^1(G)$ so $c_\alpha(f) \to 0$ as $\alpha \to \infty$ on $\hat{G}_0$.

4.10 Remark: Combining 4.7 and 4.8 with Section 3, we see that the G-operators on $L_H^1(G)$ and $C_H(G)$ are all of the form

$$f \mapsto \sum_{\alpha \in \hat{G}_0} n_\alpha\, c_\alpha(\mu) \tilde{f}_\alpha, \text{ for some } \mu \in M_{HH}(G).$$

A Schur's Lemma argument shows that any G-operator on $L_H^2(G)$ is of the form $f \mapsto \Sigma_{\alpha\in\hat{G}_0} n_\alpha s_\alpha \tilde{f}_\alpha$, $\{s_\alpha\}$ any bounded complex function on $\hat{G}_0$. There is a formal identity $\Sigma_\alpha n_\alpha s_\alpha \phi_\alpha * \Sigma_\alpha n_\alpha \tilde{f}_\alpha = \Sigma_\alpha n_\alpha s_\alpha \tilde{f}_\alpha$, which holds if the series define elements in appropriate C, L^p, or M spaces. Let (T, X) be a strongly continuous representation of G on a Banach space X, and Y a closed invariant subspace on which there is a bounded projection P, then there is a (bounded) G-projection Q on Y, that is, $Q^2 = Q$, $QX = Y$, $QT(x) = T(x)Q$ for all $x \in G$, and in fact Q is given by $\int_G T(x)PT(x^{-1})\,dx$, a strong integral (Rudin [5]). In the case $(R, L_H^2(G))$ any function $\{s_\alpha : \alpha \in \hat{G}_0, s_\alpha = 0 \text{ or } 1\}$ determines a G-projection, and an invariant subspace with a bounded projection, and conversely. For $(R, C_H(G))$ or $(R, L_H^1(G))$ the G-projections correspond to idempotent measures μ in $M_{HH}(G)$, that is $\mu * \mu = \mu$, so $c_\alpha(\mu) = 0$ or 1 for $\alpha \in \hat{G}_0$.

5. The Spaces $A_H(G)$ and $A_{HH}(G)$

5.1: The spaces $A_H(G)$ and $A_{HH}(G)$ are defined as in 2.4 and 3.3. They are both closed subalgebras of $A(G)$. Their maximal ideal spaces are G/H and $\{HxH\colon x \in G\}$ respectively. An easily accessible formula for the A-norm can be given in the case $m_\alpha = 0$ or 1 for all $\alpha \in \hat{G}$.

5.2 Lemma ([L, p. 55]): *Let A be a Banach algebra of continuous functions on a compact space X which separates points, is inverse closed, and is self-adjoint; then the maximal ideal space of A is X.*

5.3 Proposition: *The maximal ideal spaces of $A_H(G)$ and $A_{HH}(G)$ are G/H and $\{HxH : x \in G\}$ respectively.*

Proof: To use Lemma 5.2 it suffices to show that $A_H(G)$ and $A_{HH}(G)$ separate the points of G/H and $\{HxH\colon x \in G\}$ respectively. Let K be compact $\subset U$ open $\subset G/H$, then $\rho^{-1}K$ compact $\subset \rho^{-1}U$ open $\subset G$, and $f = (1/m_G(V))\,\chi_{(\rho^{-1}K)V^{-1}} * \chi_V$ is 1 on $\rho^{-1}K$, 0 off $\rho^{-1}U$, for a suitable neighborhood V of e (as in 8.5.2), and $m_H * f = f$. A similar argument works on $A_{HH}(G)$ by using a neighborhood V such that $\chi_V * m_H = \chi_V$, that is $VH = V$. The situation is K compact $\subset U$ open $\subset G/H$, with $KH = K$, $UH = U$, thus we can choose a neighborhood V of e in G such that $(\rho^{-1}K)V^{-1}V \subset \rho^{-1}U$, and $HVH = V$. $\square$

5.4: Suppose that $m_\alpha = 0$ or 1 for all $\alpha \in \hat{G}$. Then for

$$f \in A_H(G),\ \|\hat{f}_\alpha\|_1 = \left(\sum_{j=1}^{n_\alpha} |\hat{f}_{\alpha j1}|^2\right)^{1/2} = \|\hat{f}_\alpha\|_2,$$

for $\alpha \in \hat{G}_0$ and $\| f \|_A = \Sigma_{\alpha \in \hat{G}_0} n_\alpha \| \hat{f}_\alpha \|_2$. If further $f \in A_{HH}(G)$, then

$$\| \hat{f}_\alpha \|_1 = |c_\alpha(f)| \text{ for } \alpha \in \hat{G}, \text{ so } \| f \|_A = \sum_\alpha n_\alpha |c_\alpha(f)|.$$

Thus the spaces of series $\Sigma_{\alpha \in \hat{G}_0} n_\alpha \Sigma_{j=1}^{n_\alpha} \hat{f}_{\alpha j 1} T_{\alpha 1 j}$ and $\Sigma_{\alpha \in \hat{G}_0} n_\alpha c_\alpha(f)\phi_\alpha$ form Banach algebras of continuous functions under the norms

$$\sum_{\alpha \in \hat{G}_0} n_\alpha \left(\sum_{j=1}^{n_\alpha} |\hat{f}_{\alpha j 1}|^2 \right)^{1/2} \text{ and } \sum_{\alpha \in \hat{G}_0} n_\alpha |c_\alpha(f)|$$

respectively.

6. The Special Orthogonal Group and the Sphere

6.1: We will apply the preceding theory to the sphere S^{n-1}, which is the homogeneous space $\mathbf{SO}(n)/\mathbf{SO}(n-1)$. Here will be found the determination of the spherical functions, the idempotent measures in M_{HH}, the maximal ideal space of M_{HH}, and some application to harmonic functions. We fix $n \geq 3$.

$\mathbf{SO}(n)$ is the group of $n \times n$ real orthogonal matrices $g = (g_{i1})$ of determinant 1. By definition $gg^* = g^*g = I$, that is $\sum_{j=1}^n g_{ij}g_{kj} = \delta_{ik}$ and $\sum_{j=1}^n g_{ji}g_{jk} = \delta_{ik}$. The group $\mathbf{SO}(n)$ acts on $\mathbf{R}^n$ and leaves distances invariant, that is, $|xg| = |x|$ for all $x \in \mathbf{R}^n$, $|x| = (\Sigma_j x_j^2)^{1/2}$. In particular the unit sphere $S^{n-1} = \{x \in \mathbf{R}^n: |x| = 1\}$ is a homogeneous space for $\mathbf{SO}(n)$. The normalized surface integral gives the $\mathbf{SO}(n)$-invariant measure ω. To express S^{n-1} as $\mathbf{SO}(n)/H$ we choose $p = (1, 0, \dots, 0) \in S^{n-1}$ and let $H = \{g \in \mathbf{SO}(n): pg = p\}$. It is clear that $g \in H$ if and only if $g_{11} = 1$ (implying $g_{lj} = g_{jl} = 0, 2 \leq j \leq n$), and $H \cong \mathbf{SO}(n-1)$. The projection map ρ: $\mathbf{SO}(n) \to S^{n-1}$ is given by $\rho g = pg = (g_{11}, g_{12}, \dots, g_{1n})$. We will use R to denote the action of $\mathbf{SO}(n)$ on functions on $\mathbf{R}^n$ as well as the usual meaning of R.

6.2 Definition: For $m \in Z_+$ let $\mathscr{P}_m$ be the linear space of **homogeneous polynomials** on $\mathbf{R}^n$ of degree m, with complex coefficients. This space is spanned by the **monomials** $x^\alpha = x_1^{\alpha_1} x_2^{\alpha_2} \dots x_n^{\alpha_n}$, $\alpha_j \in Z_+$, $|\alpha| = \sum_{j=1}^n \alpha_j = m$. Denote the set of multi-indices $\alpha = (\alpha_1, \dots, \alpha_n)$, $\alpha_j \in Z_+$ by Z_+^n. To compute the dimension of $\mathscr{P}_m$, namely, the number of $\alpha \in Z_+^n$ such that $|\alpha| = m$, observe that the power series expansion of

$$((1 - x_1)(1 - x_2) \dots (1 - x_n))^{-1} = (1 + x_1 + x_1^2 + \cdots) \dots (1 + x_n + x_n^2 + \cdots)$$

contains each x^α exactly once, so the coefficient of t^m in $(l - t)^{-n}$ (setting

$x_1 = x_2 = \cdots = x_n = t$) is the required number; the binomial theorem shows that

$$\dim \mathscr{P}_m = \binom{n+m-1}{m} = \frac{(n+m-1)!}{m!(n-1)!}.$$

The space $\mathscr{P}_m$ is invariant under $\mathbf{SO}(n)$, that is, $f \in \mathscr{P}_m$ implies $R(g)f \in \mathscr{P}_m$ for all $g \in \mathbf{SO}(n)$.

The Laplacian

$$\Delta = \sum_{j=1}^{n} \left(\frac{\partial}{\partial x_j}\right)^2$$

is a second-order differential $\mathbf{SO}(n)$-operator, that is, $\Delta R(g)f = R(g)(\Delta f)$, for $f \in C^2(\mathbf{R}^n)$, $g \in \mathbf{SO}(n)$. For $m \in Z_+$ let $\mathscr{H}_m = \{f \in \mathscr{P}_m : \Delta f = 0\}$, the space of **harmonic homogeneous polynomials** of degree m; $\mathscr{H}_m$ is an invariant subspace of $\mathscr{P}_m$. We will show that each $\mathscr{H}_m$ is irreducible.

6.3 Theorem: *For $m \in Z_+$, each $f \in \mathscr{P}_m$ has a unique expansion $\Sigma_{j=0}^{[m/2]} |x|^{2j} f_{m-2j}$, where $f_{m-2j} \in \mathscr{H}_{m-2j}$ ($[t]$ = largest integer $\le t$). That is,*

$$\mathscr{P}_m = \mathscr{H}_m \oplus |x|^2 \mathscr{H}_{m-2} \oplus |x|^4 \mathscr{H}_{m-4} \cdots.$$

The proof proceeds in several steps.

6.3 (i): For $f(x) = \Sigma_{|\alpha|=m} a_\alpha x^\alpha \in \mathscr{P}_m$, let $\bar{f}(\partial)$ be the differential operator

$$\sum_{|\alpha|=m} \overline{a_\alpha} \partial^\alpha, \text{ where } \partial^\alpha = \left(\frac{\partial}{\partial x_1}\right)^{\alpha_1} \cdots \left(\frac{\partial}{\partial x_n}\right)^{\alpha_n}.$$

If $f \neq 0$, then $\bar{f}(\partial) f(x) > 0$. This follows since for $\beta \in Z_+^n$, $|\beta| = m$, $\partial^\alpha x^\beta = \delta_{\alpha\beta}\, \alpha!$, where $\alpha! = \alpha_1! \ldots \alpha_n!$.

6.3 (ii): Let $q \in \mathscr{P}_k$, $f \in \mathscr{P}_m$, then $\bar{q}(\partial) f \in \mathscr{P}_{m-k}$ and $\bar{q}(\partial)$ is a linear map of $\mathscr{P}_m$ into $\mathscr{P}_{m-k}$ (interpreting $\mathscr{P}_s = (0)$ if $s < 0$), thus $\dim(\ker_m \bar{q}(\partial)) \ge \dim \mathscr{P}_m - \dim \mathscr{P}_{m-k}$, where $\ker_m \bar{q}(\partial) = \{f \in \mathscr{P}_m : \bar{q}(\partial) f = 0\}$.

6.3 (iii): Let $q \in \mathscr{P}_k$, $q \neq 0$, $f \in \mathscr{P}_m$, $f \neq 0$ then $\bar{q}(\partial)\ \overline{(qf)} \neq 0$. Since the operators of the form $\bar{q}(\partial)$ all commute, we have that $\overline{(qf)}(\partial) = \bar{f}(\partial)\bar{q}(\partial)$ and by (i), $\bar{f}(\partial)\bar{q}(\partial)(qf) \neq 0$.

6.3 (iv): Let $q \in \mathscr{P}_k$, $q \neq 0$, then each $\mathscr{P}_m = \ker_m \bar{q}(\partial) \oplus q\mathscr{P}_{m-k}$. Part (iii) shows that $\ker_m \bar{q}(\partial) \cap q\mathscr{P}_{m-k} = (0)$, and part (ii) shows $\dim \ker_m \bar{q}(\partial) + \dim q\mathscr{P}_{m-k} \ge \dim \mathscr{P}_m$.

6.3 (v): Now put $q(x) = |x|^2$, then $\overline{q}(\partial) = \Delta$ and an inductive argument applied to (iv) finishes the proof. □

6.4 Corollary:

$$\dim \mathcal{H}_m = \dim \mathcal{P}_m - \dim \mathcal{P}_{m-2} = \binom{n+m-3}{m}\left(\frac{2m}{n-2}+1\right),$$

denoted by D_m^n.

6.5 Theorem: *$Sp\{\mathcal{H}_m : m \in Z_+\}$ is dense in $C(S^{n-1})$ and $L^p(S^{n-1})$, $1 \leq p < \infty$.*

Proof: By the Stone-Weierstrass theorem the space of all polynomials is dense in $C(S^{n-1})$. Each polynomial is a sum of homogeneous ones, and if $f \in \mathcal{P}_m$, then by 6.3, $f(x) = \sum_{j=0}^{[m/2]} f_{m-2j}(x)$ for all $x \in S^{n-1}$ (that is, $|x| = 1$), with $f_{m-2j} \in \mathcal{H}_{m-2j}$. □

6.6 Convention: To keep the notation neat it will be convenient to identify $C_H(\mathbf{SO}(n))$ with $C(S^{n-1})$ (and similarly for the L^p and M spaces). Observe that $f \in C_H(\mathbf{SO}(n))$ if and only if f depends only on $(g_{11}, \ldots, g_{1n}) = pg$, which is a point on S^{n-1}. We will write $f(g)$ if f is taken in the $C_H(\mathbf{SO}(n))$ context, $f(x)$ for the $C(S^{n-1})$ case.

The space $C_{HH}(\mathbf{SO}(n))$ (and similarly L^p, M) is thus identified with $\{f \in C(S^{n-1}) : R(h)f = f \text{ for all } h \in H\}$. Suppose $f \in C_{HH}(\mathbf{SO}(n))$ then there is a unique $\sigma f \in C([-1, 1])$ such that $f(g) = \sigma f(g_{11})$ or $f(x) = \sigma f(x_1)$ for all $g \in \mathbf{SO}(n)$, $x \in S^{n-1}$. For if $h_1 g_1 h_2 = g_2$, with $g_1, g_2 \in \mathbf{SO}(n)$, $h_1, h_2 \in H$ then $(g_1)_{11} = (g_2)_{11}$. Conversely if $(g_1)_{11} = (g_2)_{11}$ then there exists $h_2 \in H$ such that $(g_1 h_2)_{1j} = (g_2)_{1j}$ for $1 \leq j \leq n$, then $pg_1 h_2 = pg_2$, so $h_1 g_1 h_2 = g_2$ for some $h_1 \in H$.

6.7 Theorem: *For $m \in Z_+$, $\dim(C_{HH}(\mathbf{SO}(n)) \cap \mathcal{H}_m) = 1$, thus $\mathcal{H}_m$ is irreducible. The spherical function for $\mathcal{H}_m$ is $\phi_m(g) = P_m^{(n/2)-1}(g_{11})$ (an ultraspherical polynomial, discussed below).*

Proof: We look for $f \in \mathcal{H}_m$ such that $f | S^{n-1}$ depends only on x_1, that is, $f(x) = |x|^m \sigma f(x_1/|x|)$ for all $x \in \mathbf{R}^n$, $x \neq 0$. (Observe that if $f \in \mathcal{P}_m$, then $f(x) = |x|^m f(x/|x|)$, $x \neq 0$.) A chain rule computation shows that

$$0 = \Delta f(x) = |x|^{m-2}[(1-t^2)D^2\sigma f(t) - (n-1)tD\sigma f(t) + m(m+n-2)\sigma f(t)]$$

where $t = x_1/|x|$, and $D = d/dt$. The only polynomial solution of this is $cP_m^{n/2-1}(t)$, c constant. Thus $C_{HH}(\mathbf{SO}(n)) \cap \mathcal{H}_m$ is one-dimensional and by 4.4 and 4.5, $\mathcal{H}_m$ is irreducible with spherical function $\phi_m(g) = P_m^{(n/2)-1}(g_{11})$.

□

6.8 Note on ultraspherical polynomials: (For a reference see [AS, pp. 773–783]). The **ultraspherical polynomials** P_m^λ of degree m, index $\lambda > 0$ are given by

$$(1 - 2rt + r^2)^{-\lambda} = \sum_{m=0}^{\infty} \frac{\Gamma(2\lambda + m)}{\Gamma(2\lambda)m!} P_m^\lambda(t) r^m.$$

Then $P_m^\lambda(1) = 1$, $P_m^\lambda(-1) = (-1)^m$, and P_m^λ is an even or odd polynomial according to m being even or odd respectively. The class $\{P_m^\lambda(t): m \in Z_+\}$ is the family of orthogonal polynomials for the weight function $(1 - t^2)^{\lambda - 1/2}\, dt$ on $[-1, 1]$. Further $P_m^\lambda(t)$ is the unique polynomial solution of

$$[(1 - t^2)D^2 - (2\lambda + 1)tD + m(m + 2\lambda)]f(t) = 0.$$

6.9 Theorem: *For each* $\alpha \in (\mathbf{SO}(n))\hat{}$, $m_\alpha = 0$ *or* 1 *and* $L_H^2(\mathbf{SO}(n)) = \Sigma \oplus_{m=0}^{\infty} \mathscr{H}_m$, *that is, the irreducible unitary continuous representations of* $\mathbf{SO}(n)$ *which contain* $H \to \{1\}$ *are exactly the classes of* $(R, \mathscr{H}_m)$, $m \in Z_+$, *and each* $(R, \mathscr{H}_m)$ *contains* $H \to \{1\}$ *just once.*

Proof: By 6.7 and 4.4 each $(R, \mathscr{H}_m)$ contains $H \to \{1\}$ just once. For $m \neq k$, $\mathscr{H}_m$ is not equivalent to $\mathscr{H}_k$, since $\phi_m \neq \phi_k$, thus $\mathscr{H}_m \perp \mathscr{H}_k$ (in $L^2(\mathbf{SO}(n))$ or $L^2(S^{n-1})$). By 6.5, $Sp\{\mathscr{H}_m: m \in Z_+\}$ is dense in $L^2(S^{n-1})$ or $L_H^2(\mathbf{SO}(n))$, thus $L^2(S^{n-1}) = \Sigma \oplus_{m=0}^{\infty} \mathscr{H}_m$. But in general $L_H^2(G) = \Sigma \oplus \{\phi_\alpha * C_H(G): \alpha \in \hat{G}, m_\alpha > 0\}$, so the proof is done. □

6.10 Convolutions on $\mathbf{S}^{n-1}$: Let $f \in C_{HH}(\mathbf{SO}(n))$, then $f(g) = \sigma f(g_{11})$, for some $\sigma f \in C([-1, 1])$ (see 6.6). A computation with spherical polar coordinates shows that

$$\int_{\mathbf{SO}(n)} f(g)\, dg = a_n \int_{-1}^{1} \sigma f(t)\, (1 - t^2)^{(n-3)/2}\, dt$$

where $a_n = (\int_{-1}^{1} (1 - t^2)^{(n-3)/2}\, dt)^{-1}$, thus σ extends to an isomorphism of $L_{HH}^1(\mathbf{SO}(n))$ onto $L^1([-1, 1], a_n(1 - t^2)^{(n-3)/2}\, dt)$. Now let $f \in L_{HH}^1(\mathbf{SO}(n))$, then by 4.8, f has the Fourier series

$$\sum_{m=0}^{\infty} D_m^n c_m(f) \phi_m,$$

where $c_m(f) = \int_{\mathbf{SO}(n)} f(g)\phi_m(g)\, dg = a_n \int_{-1}^{1} \sigma f(t) P_m^{n/2-1}(t)\, (1 - t^2)^{(n-3)/2}\, dt$, and the maps $\{c_m\}$ form the maximal ideal space of $L_{HH}^1(\mathbf{SO}(n))$. Now suppose $q \in L^1(S^{n-1})$, then $f * q \in L_H^1(\mathbf{SO}(n))$, and

$$f * q(g_0) = \int_{\mathbf{SO}(n)} f(g_0 g^{-1}) q(pg)\, dg$$

$$= \int_{\mathbf{SO}(n)} \sigma f\left(\sum_{j=1}^{n} g_{01j} g_{1j}\right) q(pg)\, dg = \int_{S^{n-1}} \sigma f\left(\sum_{j=1}^{n} g_{01j} x_j\right) q(x)\, d\omega(x)$$

for almost all $g_0 \in \mathbf{SO}(n)$. Now put $y = pg_0$, and obtain

$$f * q(y) = \int_{S^{n-1}} \sigma f(x \cdot y) q(x)\, d\omega(x),$$

for almost all $y \in S^{n-1}$, where

$$x \cdot y = \sum_{j=1}^{n} x_j y_j.$$

Each $q \in L^1(S^{n-1})$ has a Fourier series

$$\sum_{m=0}^{\infty} D_m^n \phi_m * q,$$

then

$$\tilde{q}_m(x) = \phi_m * q(x) = \int_{S^{n-1}} P_m^{n/2-1}(x \cdot y) q(y)\, d\omega(y),$$

for all $x \in S^{n-1}$ and $\tilde{q}_m \in \mathscr{H}_m$. This is the **Laplace series** for functions on S^{n-1},

$$q \sim \sum_{m=0}^{\infty} D_m^n \tilde{q}_m, \qquad \tilde{q}_m \in \mathscr{H}_m.$$

Proposition 4.7 shows that if $f \in L^1_{HH}(\mathbf{SO}(n))$ and $q \in L^1(S^{n-1})$ then $f * q$ has the Laplace series

$$\sum_{m=0}^{\infty} D_m^n c_m(f) \tilde{q}_m.$$

6.11 The Poisson integral: Suppose f is a harmonic polynomial on $\mathbf{R}^n$, then

$$f = \sum_{m=0}^{\infty} D_m^n \tilde{f}_m,$$

where $\tilde{f}_m \in \mathscr{H}_m$, (a finite sum). We can reconstruct f in $\{|x| < 1\}$ from its values on S^{n-1}, since

$$f(x) = \sum_{m=0}^{\infty} D_m^n \tilde{f}_m(x) = \sum_{m=0}^{\infty} D_m^n |x|^m \tilde{f}_m(x/|x|)$$

$$= \left(\sum_m D_m^n |x|^m \phi_m \right) * \left(\sum_m D_m^n \tilde{f}_m \right)(x/|x|)$$

formally. That is for $0 < |x| < 1$, $f(x) = P_{|x|} * f(x/|x|)$ where

$$P_r = \sum_{m=0}^{\infty} D_m^n r^m \phi_m, \qquad 0 \le r < 1.$$

Since $D_m^n = O(m^{n-2})$ we see that each $P_r \in A_{HH}(\mathbf{SO}(n))$, $0 \leq r < 1$. To compute P_r we observe that the generating function in 6.8 can be written as

$$(1 - 2rt + r^2)^{-\lambda} = \sum_{m=0}^{\infty} D_m^n \left(\frac{m}{\lambda} + 1\right)^{-1} P_m^\lambda(t) r^m, \qquad 0 \leq r < 1$$

where $\lambda = (n/2) - 1$, and we differentiate it with respect to r, multiply by r/λ, and add it to the original, obtaining $\sigma P_r(t) = (1 - r^2)(1 - 2rt + r^2)^{-n/2}$. Then

$$P_r \geq 0, \qquad \int_{\mathbf{SO}(n)} P_r(g)\, dg = \int_{\mathbf{SO}(n)} \sum_{m=0}^{\infty} D_m^n r^m \phi_m(g)\, dg = 1;$$

and for a fixed $s < 1$, $P_r(g) \leq \sigma P_r(s)$ if $g_{11} \leq s$, and $\sigma P_r(s) \to 0$ as $r \to 1$.

6.12 Definition: For $f \in L^1(S^{n-1})$, $\mu \in M(S^{n-1})$, define the **Poisson integral** in $\{|x| < 1\}$ by $P[f](x) = P_{|x|} * f(x/|x|)$, $x \neq 0$

$$P[f](0) = P_0 * f = \int_{S^{n-1}} f\, d\omega,$$

and similarly

$$P[d\mu](x) = P_{|x|} * \mu(x/|x|).$$

6.13 Theorem: *For $f \in L^1(S^{n-1})$, $\mu \in M(S^{n-1})$, $P[f]$ and $P[d\mu]$ are continuous and harmonic in $\{|x| < 1\}$. Furthermore*

(i) *if $f \in C(S^{n-1})$ then $\| P_r * f - f \|_\infty \to 0$ as $r \to 1$*

(ii) *if $f \in L^p(S^{n-1})$, $1 \leq p < \infty$, then $\| P_r * f - f \|_p \to 0$ as $r \to 1$*

(iii) *$P_r * \mu \to \mu$ in the weak-$*$ topology as $r \to 1$*

(iv) *if $f \in C(S^{n-1})$, then the function g on $\{|x| \leq 1\}$ defined by*

$$g(x) = \begin{cases} f(x) & |x| = 1 \\ P[f](x) & |x| < 1 \end{cases}$$

is continuous.

Proof: The series for $P[f](x)$ converges uniformly in $|x| \leq r < 1$, so Harnack's theorem shows it is harmonic.

Parts (i)–(iii) are a standard approximate identity property; and we sketch this technique for homogeneous spaces. For $f \in C_H(\mathbf{SO}(n))$, $\varepsilon > 0$, there exists a neighborhood U of e such that $\| L(g)f - f \|_\infty < \varepsilon$ for all $g \in U$, and in fact for all $g \in H\,U\,H$. Let $g = h_1 g_1 h_2$, $h_1, h_2 \in H$, $g_1 \in U$, then

$$|L(g)f(g_0) - f(g_0)| = |f(h_2^{-1} g_1^{-1} h_1^{-1} g_0) - f(g_0)|$$
$$= |f(g_1^{-1} h_1^{-1} g_0) - f(h_1^{-1} g_0)| = |L(g_1^{-1}) f(h_1^{-1} g_0) - f(h_1^{-1} g_0)| < \varepsilon,$$

for all $g_0 \in \mathbf{SO}(n)$. But a connected neighborhood of the form $H\,U\,H$ is exactly a set $\{g \in \mathbf{SO}(n): g_{11} > 1 - \delta\}$, so let $f \in C_H(\mathbf{SO}(n))$, $g_0 \in \mathbf{SO}(n)$, then

$$P_r * f(g_0) - f(g_0) = \int_{\mathrm{SO}(n)} P_r(g)\,(f(g^{-1}g_0) - f(g_0))\,dg$$

$$= \left\{\int_{g_{11} > 1-\delta} + \int_{g_{11} \le 1-\delta}\right\} \sigma P_r(g_{11})\,(L(g)f(g_0) - f(g_0))\,dg.$$

Now the first integral can be made small by making δ small, since

$$L(g)f(g_0) - f(g_0) \to 0 \quad \text{as} \quad g_{11} \to 1,$$

and the second integral tends to zero as $r \to 1$. A similar argument works for $L^p_H(\mathbf{SO}(n))$. □

6.14 Theorem: *Let $\mu \in M(S^{n-1})$ with $d\mu = f\,d\omega + d\mu_s$, where $f \in L^1(S^{n-1})$ and $\mu_s \perp \omega$, then $P[d\mu](rx) \to f(x)$ as $r \to 1$ for ω-almost all $x \in S^{n-1}$.*

Proof: The idea is to use set-theoretic differentiation on μ, that is, put

$$D\mu(x) = \lim_{s \to 1} \frac{\mu\{y : y \cdot x > s\}}{\omega\{y : y \cdot x > s\}},$$

where it exists ($x, y \in S^{n-1}$). The techniques of differentiation [RRC, p. 154] using the local homeomorphism of S^{n-1} into $\mathbf{R}^{n-1}$, show that $D\mu(x)$ exists and equals $f(x)$ ω-almost everywhere. The rest is as in [RRC, pp. 225–6]. □

6.15 Notation: Let $M'_{HH}(\mathbf{SO}(n)) = \{\mu \in M_{HH}(\mathbf{SO}(n)) : |\mu|\{p, -p\} = 0\}$ (where we consider μ as a measure on S^{n-1}, see 6.6), then $M'_{HH}(\mathbf{SO}(n))$ is exactly the set of continuous measures in $M_{HH}(\mathbf{SO}(n))$ since the only countable set in S^{n-1} invariant under H is $\{p, -p\}$.

6.16 Theorem: *Suppose $\mu, \nu \in M'_{HH}(\mathbf{SO}(n))$, then $\mu * \nu \ll \omega$, that is, $d(\mu * \nu) = f\,d\omega$, for some $f \in L^1_{HH}(\mathbf{SO}(n))$.*

Proof: (i) If $\mu \in M_{HH}(\mathbf{SO}(n))$ has the property that $\mu K = 0$ whenever K is compact $\subset S^{n-1}$ and $KH = K$, and $\omega(K) = 0$, then $\mu \ll \omega$.

(ii) For such a set K, $\chi_K \in L^1_{HH}(\mathbf{SO}(n))$ and $\sigma\chi_K(t) = 0$ dt-almost everywhere, since $\omega K = a_n \int_{-1}^{1} \sigma\chi_K(t)\,(1 - t^2)^{(n-3)/2}\,dt$.

(iii) For a compact $K \subset S^{n-1}$ such that $KH = K$, and $\mu, \nu \in M_{HH}(\mathbf{SO}(n))$ we have

$$\mu * \nu(K) = \int_{S^{n-1}} d\mu(x) \int_{S^{n-1}} d\nu(y) \int_H \sigma\chi_K(xh \cdot y)\,dm_H(h).$$

This formula is derived as follows: For $f \in C(S^{n-1})$, $\mu \in M_{HH}(\mathbf{SO}(n))$

$$\int_{S^{n-1}} f(x)\,d\mu(x) = \int_{S^{n-1}} d\mu(x) \int_H f(xh)\,dm_H(h),$$

and for $f \in C_{HH}(\mathbf{SO}(n))$, $\mu, \nu \in M_{HH}(\mathbf{SO}(n))$,

$$\int_{\mathbf{SO}(n)} f\, d(\mu * \nu) = \int_{\mathbf{SO}(n)} \int_{\mathbf{SO}(n)} f(gg')\, d\mu(g)\, d\nu(g')$$

$$= \int_{\mathbf{SO}(n)} \int_{\mathbf{SO}(n)} \sigma f(pg \cdot pg')\, d\mu(g)\, d\check{\nu}(g')$$

(but $\check{\nu} = \nu$, since

$$c_m(\check{\nu}) = \int \phi_m(g^{-1})\, d\nu(g) = \int \phi_m(g)\, d\nu(g) = c_m(\nu)$$

for $m \in Z_+$)

$$= \int_{S^{n-1}} d\nu(y) \int_{S^{n-1}} d\mu(x) \int_H \sigma f(xh \cdot y)\, dm_H(h).$$

Now we extend this formula to χ_K by means of the dominated convergence theorem.

(iv) $\int_H \sigma\chi_K(xh \cdot y)\, dm_H(h)$

$$= c \int_{-1}^{1} \sigma\chi_K(st + u((1 - s^2)(1 - t^2))^{1/2})(1 - u^2)^{(n-4)/2}\, du$$

where $s = x_1$, $t = y_1$, and c is a constant depending only on n. This formula is derived by changing the integral over H to one over $E = \{z \in S^{n-1} : z_1 = x_1\}$ and using a spherical polar coordinate system for E.

(v) The above integral vanishes if $s, t \notin \{-1, 1\}$ and $\sigma\chi_K = 0$ a.e.

(vi) If $\mu, \nu \in M'_{HH}(\mathbf{SO}(n))$ and K is as in (i) then $\mu * \nu(K) = 0$ since $s, t = 1$ or -1 only for $x, y = p$ or $-p$ where both μ and ν have no mass. Thus $\mu * \nu \ll \omega$. □

6.17 Theorem: *Define bounded linear functionals c_+ and c_- on $M_{HH}(\mathbf{SO}(n))$ by $c_+(\mu) = \mu\{p\} + \mu\{-p\}$, and $c_-(\mu) = \mu\{p\} - \mu\{-p\}$,*

then

$$\lim_{m \to \infty} c_{2m}(\mu) = c_+(\mu)$$

and

$$\lim_{m \to \infty} c_{2m+1}(\mu) = c_-(\mu)$$

for all $\mu \in M_{HH}(\mathbf{SO}(n))$, thus c_+ and c_- are multiplicative. Further the maximal ideal space of $M_{HH}(\mathbf{SO}(n))$ is $\{c_m : m \in Z_+\} \cup \{c_+, c_-\}$.

Proof: Let F be a nonzero multiplicative linear functional on $M_{HH}(\mathbf{SO}(n))$, then either $F(\mu) \neq 0$ for some $\mu \in M'_{HH}(\mathbf{SO}(n))$ or $F(M'_{HH}(\mathbf{SO}(n))) = 0$. In the first case $F(\mu * \mu) \neq 0$, but $\mu * \mu \in L'_{HH}(\mathbf{SO}(n))$, so $F(f) = c_m(f)$ for some $m \in Z_+$(6.10 and 4.9) for all $f \in L'_{HH}(\mathbf{SO}(n))$. Now let $f\,d\omega = d(\mu * \mu)$ so $F(f) \neq 0$, and let $\nu \in M_{HH}(\mathbf{SO}(n))$ then $f * \nu \in L'_{HH}(\mathbf{SO}(n))$ and

$$c_m(f)c_m(\nu) = c_m(f * \nu) = F(f * \nu) = F(f)F(\nu) = c_m(f)F(\nu),$$

thus $F(\nu) = c_m(\nu)$, $F = c_m$, since $c_m(f) \neq 0$.

Otherwise $F(M'_{HH}(\mathbf{SO}(n))) = 0$, and any $\mu \in M_{HH}(\mathbf{SO}(n))$ decomposes uniquely, $\mu = \mu_0 + a\delta_p + b\delta_{-p}$, where $\mu_0 \in M'_{HH}(\mathbf{SO}(n))$, $a = \mu\{p\}$, $b = \mu\{-p\}$, δ_x is the unit mass at $x \in S^{n-1}$, thus $F(\mu) = aF(\delta_p) + bF(\delta_{-p})$. But δ_p is the identity in $M_{HH}(\mathbf{SO}(n))$ (namely m_H) so $F(\delta_p) = 1$. Further $\delta_{-p} * \delta_{-p} = \delta_p$ so $F(\delta_{-p}) = \pm 1$, (note $c_m(\delta_{-p}) = P_m^{n/2-1}(-1) = (-1)^m$, so $c_m(\delta_{-p} * \delta_{-p}) = 1$ all m). Now $\mu_0 * \mu_0 \in L'_{HH}(\mathbf{SO}(n))$, thus $c_m(\mu_0 * \mu_0) \to 0$ as $m \to \infty$, so $c_m(\mu_0) \to 0$. Finally $c_{2m}(\mu) = c_{2m}(\mu_0) + a + b \to a + b = c_+(\mu)$ as $m \to \infty$, $c_{2m+1}(\mu) = c_{2m+1}(\mu_0) + a - b \to a - b = c_-(\mu)$ as $m \to \infty$. □

6.18: Let $\mu_1 = \frac{1}{2}\delta_p + \frac{1}{2}\delta_{-p}$, $\mu_2 = \frac{1}{2}\delta_p - \frac{1}{2}\delta_{-p}$, then $c_m(\mu_1) = \frac{1}{2}(1 + (-1)^m)$ and $c_m(\mu_2) = \frac{1}{2}(1 - (-1)^m)$ thus μ_1, μ_2, δ_p are idempotent measures. If μ is any **idempotent** in $M_{HH}(\mathbf{SO}(n))$ then, since $\lim_{m\to\infty} c_{2m}(\mu)$ and $\lim_{m\to\infty} c_{2m+1}(\mu)$ have to exist, this means that $\{c_m(\mu)\}$ has period 2 except for a finite number of entries. The finite number of entries are changed by adding suitable finite sums $\Sigma_m D_m^n a_m \phi_m$ to one of 0, μ_1, μ_2, δ_p $(a_m = 0, \pm 1)$.

6.19 Theorem: *Let $E \subset Z_+$, then the following are equivalent:*

(i) *For each complex function F on E tending to 0 at ∞ there exists*

$$f \in L^1_{HH}(\mathbf{SO}(n))$$

such that $c_m(f) = F(m)$ for all $m \in E$.

(ii) *For each bounded complex function F on E there exists $\mu \in M_{HH}(\mathbf{SO}(n))$ such that $c_m(\mu) = F(m)$ for all $m \in E$.*

(iii) *E is finite.*

Proof: The proof that (i) implies (ii) is the usual Sidon set argument using the closed graph theorem and the weak-$*$ compactness of bounded sets in $M_{HH}(\mathbf{SO}(n))$ (see [R, p. 121]). For (ii) implies (iii) define F by $F(4k) = F(4k+1) = 1$, $F(4k+2) = F(4k+3) = 0$, $k \in Z_+$ and restrict it to E. Let $\mu \in M_{HH}(\mathbf{SO}(n))$ such that $c_m(\mu) = F(m)$ for all $m \in E$. But then $\lim_{m\to\infty} c_{2m+1}(\mu)$ and $\lim_{m\to\infty} c_{2m}(\mu)$ exist so E must be finite. For (iii) implies (i) put $f = \Sigma_m D_m^n F(m)\phi_m$. □

7. Historical Notes

7.1: Theorem 3.6 is an adaptation to homogeneous spaces of a result of Helgason [1]. The theory of spherical functions is presented in more generality by Godement [1]; in particular, he discovered Theorem 4.8. There is a discussion of modules over $A_H(G)$ in Dunkl [4]. Note that 5.4 applied to S^{n-1} shows that the space of series $f(x) = \sum_{m=0}^{\infty} D_m^n \tilde{f}_m(x)$ forms a Banach algebra of continuous functions on S^{n-1} under the norm

$$\| f \|_A = \sum_{m=0}^{\infty} (D_m^n)^{3/2} \| \tilde{f}_m \|_2.$$

This was noticed by Rider [1]. The presentation of Theorem 6.3 on spherical harmonics is taken from Calderón [C, pp. 29–30]. The spherical functions for S^{n-1} were discovered by É. Cartan [1]. Theorems 6.16–6.19 are from Dunkl [1].

An application of convolution on S^{n-1} to determine the polynomial $q_\alpha \in \mathscr{P}_m$, given $\alpha \in Z_+^n$, $|\alpha| = m$, such that

$$\int_{S^{n-1}} q_\alpha(x) \sum_{|\beta| = m} a_\beta x^\beta \, d\omega(x) = a_\alpha$$

is to be found in Dunkl [2].

CHAPTER 10

ANALYTIC FUNCTIONS ON THE BALL

1. Introduction

1.1: The space of analytic functions on the ball $\{z:\sum_{j=1}^{n} |z_j|^2 < 1\}$ in $\mathbf{C}^n$ is invariant under the action of the unitary group. This fact suggests the application of compact group theory to these analytic functions. With the machinery on hand, we can derive the Cauchy integral formula, which reproduces certain analytic functions from their boundary values, and the H^p-theory for the ball. However, as Rudin [RFT] points out, results in several variables analogous to the one-variable results are not as exciting as the results peculiar to several variables. We thus present the H^p-theory for the sake of reference for further work in the area. We will refer to [RFT] for the proofs in the cases where the polydisc proof works with but a change in name. The one several-variables result deals with the space of analytic functions continuous on the closed ball and is to be found in Section 4.14.

2. Analytic Functions and the Unitary Group

2.1 Notation: $\mathbf{C}^n$ is the space $\{z:z = (z_1, \ldots, z_n), z_j \in \mathbf{C}\}$ with the Euclidean norm $|z| = (\sum_{j=1}^{n} |z_j|^2)^{1/2}$ and inner product $\langle z, w\rangle = \sum_{j=1}^{n} z_j\bar{w}_j$. For $m \in Z_+$ let $\mathscr{P}_m$ be the space of homogeneous polynomials of degree in z, so each $f \in \mathscr{P}_m$ is a sum $f(z) = \Sigma_{|\alpha| = m} a_\alpha z^\alpha$, $\alpha \in Z_+^n$, each $a_\alpha \in \mathbf{C}$. As seen in 9.6.2

$$\dim \mathscr{P}_m = \binom{n+m-1}{m}.$$

Let Ω be a connected open set in $\mathbf{C}^n$, then $H(\Omega)$ is the space of functions analytic on Ω (continuous in z, analytic in each z_j), and $A(\Omega)$ is the space of functions analytic on Ω and continuous on $\bar{\Omega}$. Define the ball $B_n = \{z: |z| < 1\}$ and its boundary $\partial B_n = S = \{z: |z| = 1\}$. As usual, $U = \{\lambda \in \mathbf{C}: |\lambda| < 1\}$.

$\mathbf{U}(n)$ is the **unitary group** on $\mathbf{C}^n$; its elements are $n \times n$ complex matrices $u = (u_{jk})$ such that $uu^* = u^*u = I$, that is, $\sum_{k=1}^n u_{jk}\overline{u_{mk}} = \delta_{jm}$, and $\mathbf{U}(n)$ is a compact Lie group. Both B_n and S are invariant under the action $z \mapsto zu$, $u \in \mathbf{U}(n)$, and, further, S is a homogeneous space for $\mathbf{U}(n)$. Setting $p = (1, 0, \ldots, 0) \in S$, we obtain $S \cong \mathbf{U}(n)/H$ where $H = \{u \in \mathbf{U}(n): u_{11} = 1\} \cong \mathbf{U}(n-1)$.

We will again adopt a convention like 9.6.6 and identify $C(S)$ with $C_H(\mathbf{U}(n))$, and we observe that for $u \in \mathbf{U}(n)$, $pu = (u_{11}, u_{12}, \ldots, u_{1n}) \in S$. For functions f on $\mathbf{C}^n$, $u \in \mathbf{U}(n)$, let $R(u)f(z) = f(zu)$.

2.2: The space $\mathscr{P}_m$ is invariant under R. As in 9.6.6 if $f \in C_{HH}(\mathbf{U}(n))$ then f is a function of u_{11}, $u \in \mathbf{U}(n)$. Suppose $f \in \mathscr{P}_m \cap C_{HH}(\mathbf{U}(n))$, then $f(z) = |z|^m \sigma f(z_1/|z|)$, $z \neq 0$, some function σf on $\bar{U} = \{\lambda \in \mathbf{C}: |\lambda| \leq 1\}$. But for f to be a polynomial we must have $f(z) = cz_1^m$, a constant c, thus by Lemma 9.4.4, $\mathscr{P}_m$ is irreducible with the spherical function $u \mapsto u_{11}^m$.

2.3: There is a natural embedding of $\mathbf{T}^n$ in $\mathbf{U}(n)$, namely, the identification of $(e^{i\theta_1}, \ldots, e^{i\theta_n})$ with the diagonal matrix with the same elements down the diagonal. Observe that B_n is a Reinhardt domain in $\mathbf{C}^n$, that is $B_n u = B_n$ for all $u \in \mathbf{T}^n$ and B_n is open. The fundamental theorem on power series [Hö, 2.4.5, p. 35] shows that each $f \in H(B_n)$ has a power series

$$\sum_{\alpha \in Z_+^n} a_\alpha z^\alpha,$$

with normal convergence to f in B_n, that is, for each K compact $\subset B_n$,

$$\sum_\alpha \sup_{z \in K} |a_\alpha z^\alpha| < \infty,$$

so the partial sums converge to f in the compact-open topology.

2.4: There is an identification of $\mathbf{C}^n$ with $\mathbf{R}^{2n}$, where z corresponds to $(x_1, y_1, x_2, y_2, \ldots)$, $z_j = x_j + iy_j$. Then S is identified with the unit sphere in $\mathbf{R}^{2n}$, B_n with its interior. The group $\mathbf{U}(n)$ is mapped onto a subgroup of

$\mathbf{SO}(2n)$, where in the matrix (u_{jk}) the element u_{jk} is replaced by

$$\begin{pmatrix} \operatorname{Re} u_{jk} & \operatorname{Im} u_{jk} \\ -\operatorname{Im} u_{jk} & \operatorname{Re} u_{jk} \end{pmatrix}.$$

We have defined the $\mathbf{SO}(2n)$-invariant measure ω on S so it is also the unique (normalized) $\mathbf{U}(n)$-invariant measure on S. The Laplacian Δ on $\mathbf{R}^{2n}$ or $\mathbf{C}^n$ is

$$\sum_{j=1}^{n} \left(\left(\frac{\partial}{\partial x_j} \right)^2 + \left(\frac{\partial}{\partial y_j} \right)^2 \right),$$

and each $f \in \mathscr{P}_m$, $m \in Z_+$, is harmonic since analytic functions satisfy

$$\left(\left(\frac{\partial}{\partial x_j} \right)^2 + \left(\frac{\partial}{\partial y_j} \right)^2 \right) f(z) = 0$$

for each j. Thus $\mathscr{P}_m$ consists of harmonic homogeneous polynomials when viewed on $\mathbf{R}^{2n}$, so $\mathscr{P}_m \subset \mathscr{H}_m^R$ where $\mathscr{H}_m^R$ is the space of harmonic homogeneous polynomials of degree m on $\mathbf{R}^{2n}$ (we use the superscript ^{R} to emphasize the domain). Of course a polynomial on $\mathbf{R}^{2n}$ is just a polynomial in z, $\bar{z}$. Since $\mathbf{U}(n)$ is essentially a subgroup of $\mathbf{SO}(2n)$, each $\mathscr{H}_m^R$ is invariant under $\mathbf{U}(n)$, and $\mathscr{P}_m$ is an irreducible subspace of it. We can see that if f is analytic on a neighborhood of $\bar{B}_n$ then f is orthogonal (in $L^2(S)$) to the orthogonal complement of $\mathscr{P}_m$ in $\mathscr{H}_m^R$, each $m \in Z_+$. We will give further details on this later.

2.5 Remark on representations of U(n): The dual of $\mathbf{U}(n)$ is indexed by the set of weights $(m_1, \ldots, m_n)$ where each $m_j \in Z$ and $m_1 \geq m_2 \ldots m_{n-1} \geq m_n$, that is to each equivalence class of irreducible unitary continuous representations there corresponds a weight, and conversely. The class conjugate to $(m_1, \ldots, m_n)$ is $(-m_n, \ldots, -m_1)$. The representation $(R, \mathscr{H}_m^R)$ of $\mathbf{SO}(2n)$ when restricted to $\mathbf{U}(n)$ decomposes into $\sum \oplus_{j=0}^{m} (m-j, 0, \ldots 0, -j)$, in particular, $(m, 0, \ldots, 0)$ is the one realized on $\mathscr{P}_m$ and $(0, \ldots, 0, -m)$ the one on $\bar{\mathscr{P}}_m$ (polynomials in $\bar{z}$). We will use (m_1, m_n) as a short form for $(m_1, 0, \ldots, 0, m_n)$ and $D_{(m_1, m_n)}$ for the dimension of (m_1, m_n). The book of Boerner [B, Ch. V] is a reference for these statements.

2.6: We know that $Sp\{\mathscr{H}_m^R : m \in Z_+\}$ is dense in $C(S)$, and each polynomial in z is in $Sp\{\mathscr{P}_m : m \in Z_+\}$ in $C(S)$. Now let $f \in C(S)$ then it has the $\mathbf{U}(n)$-Fourier series $\sum_{m=0}^{\infty} \sum_{j=0}^{m} D_{(m-j,-j)} \phi_{(m-j,-j)} * f$ (note that this is the $\mathbf{U}(n)$-convolution over S). For our purposes it is enough to know that

$$\phi_{(m,0)}(u) = u_{11}^m \quad \text{and} \quad D_{(m,0)} = \binom{n+m-1}{m} \qquad \text{for } m \in Z_+.$$

The remaining spherical functions are expressed in terms of Jacobi polynomials.

If $f \in C(S)$ and is a polynomial in z then it has the series $\sum_{m=0}^{\infty} D_{(m,0)} \phi_{(m,0)} * f$ (finite sum) and this is the decomposition of f into homogeneous polynomials. Now for $m \in Z_+$, $f \in C(S)$, $\sum_{j=0}^{m} D_{(m-j,-j)}\, \phi_{(m-j,-j)} * f$ is a harmonic homogeneous polynomial on $\mathbf{R}^{2n}$ restricted to S, and it is extended to B_n by

$$|z|^m \sum_{j=0}^{m} D_{(m-j,-j)}\, \phi_{(m-j,-j)} * f(z/|z|),$$

$z \neq 0$, thus the Poisson integral of $f \in C(S)$ is

$$P[f](z) = \sum_{m=0}^{\infty} |z|^m \sum_{j=0}^{m} D_{(m-j,-j)}\, \phi_{(m-j,-j)} * f(z/|z|),$$

for $0 < |z| < 1$ (referring to the Poisson integral on $S^{2n-1} \subset \mathbf{R}^{2n}$). These remarks prove the following theorem.

2.7 Theorem: *Let $f \in C(S)$, then*
(i) *$P[f] \in A(B_n)$ if and only if $\phi_{(m-j,-j)} * f = 0$ for $m \geq 1$, $1 \leq j \leq m$.*
(ii) *$P[f] = \operatorname{Re} g$, where $g \in H(B_n)$ and $\operatorname{Re} g \in C(\overline{B}_n)$ if and only if $\phi_{(m-j,-j)} * f = 0$ for $m \geq 2$, $1 \leq j \leq m-1$.*

This is analogous to the polydisc theorems [RFT, p. 21, 2.1.4 and p. 33, 2.4.1].

2.8 The Cauchy integral formula: For $f \in A(B_n)$ we know $f \sim \sum_{m=0}^{\infty} D_{(m,0)}^{n}\, \phi_{(m,0)} * f$ on S since f is a uniform limit of polynomials in z. To reproduce f from its boundary values we note that $f(z) = \sum_{m=0}^{\infty} D_{(m,0)} |z|^m \phi_{(m,0)} * f(z/|z|)$ for $0 < |z| < 1$ which by 9.4.10 is $(\sum_{m=0}^{\infty} D_{(m,0)} |z|^m \phi_{(m,0)}) * f(z/|z|)$. In fact

$$\sum_{m=0}^{\infty} D_{(m,0)} |z|^m \phi_{(m,0)}(u) = (1 - |z| u_{11})^{-n}$$

for $|z| < 1$ and $u \in \mathbf{U}(n)$. Put $C_r(u) = (1 - r u_{11})^{-n}$ for $0 \leq r < 1$ then $C_r \in C_{HH}(\mathbf{U}(n))$ and

$$\begin{aligned} C_r * f(u_0) &= \int_{\mathbf{U}(n)} (1 - r(u_0 u^{-1})_{11})^{-n} f(pu)\, du \\ &= \int_{\mathbf{U}(n)} \Big(1 - r \sum_{j=1}^{n} u_{01j} \bar{u}_{1j}\Big)^{-n} f(pu)\, du, \end{aligned}$$

and transferring this to S with $z = pu_0$, $w = pu$,

$$C_r * f(z) = \int_S (1 - r\langle z, w\rangle)^{-n} f(w)\, d\omega(w)$$

for $z \in S$. We obtain the **Cauchy formula** for $f \in A(B_n)$, $z \in B_n$:

$$f(z) = C_{|z|} * f(z/|z|) = \int_S (1 - \langle z, w\rangle)^{-n} f(w)\, d\omega(w).$$

This formula is a special case of the Bochner–Martinelli formula and was derived in another way by Bungart [1].

2.9 Proposition: *Let $\alpha, \beta \in Z^n_+$, then*

$$\int_S z^\alpha \bar{z}^\beta \, d\omega(z) = \delta_{\alpha\beta} \frac{\alpha!(n-1)!}{(n+|\alpha|-1)!}$$

where $\alpha! = \alpha_1! \cdots \alpha_n!$.

Proof: Fix $\alpha \in Z^n_+$, then by Cauchy's formula

$$\int_S (1 - r\langle z, w\rangle)^{-n} w^\alpha \, d\omega(w) = r^{|\alpha|} z^\alpha, \text{ for } 0 \le r < 1, z \in S.$$

Expanding $(1 - r\langle z, w\rangle)^{-n}$ by the binomial and multinomial theorems we obtain

$$(1 - r\langle z, w\rangle)^{-n} = \sum_{m=0}^{\infty} \binom{n+m-1}{m} r^m \sum_{|\beta|=m} \binom{m}{\beta} z^\beta \bar{w}^\beta$$

$\left(\text{where } \binom{m}{\beta} = \dfrac{m!}{\beta!}\right)$. Fubini's theorem shows that

$$r^{|\alpha|} z^\alpha = \sum_{m=0}^{\infty} \binom{n+m-1}{m} r^m \sum_{|\beta|=m} \binom{m}{\beta} z^\beta \int_S \bar{w}^\beta w^\alpha \, d\omega(w).$$

By comparing coefficients of $r^{|\beta|} z^\beta$ on both sides we obtain the required formula. □

3. Subharmonic Functions

3.1: By applying the theory of subharmonic functions and a theorem of Gårding and Hörmander [1] we can prove the standard H^p-type theorems for $H(B_n)$. We will quickly sketch the subharmonic function theory indicating only the small changes that need to be made in the $\mathbf{R}^2$-proof (found in Radó [Ra], and [Hö]).

3.2 Definition: Let Ω open $\subset \mathbf{C}^n$, then g is **subharmonic** on Ω if

(i) $-\infty \le g(z) < \infty, z \in \Omega$.

(ii) g is upper semicontinuous on Ω, (that is, $\{z\colon g(z) < a\}$ is open for each $a \in \mathbf{R}$).

(iii) for K compact $\subset \Omega$, $f \in C(K)$, f harmonic in int K, and $f \ge g$ on ∂K then $f \ge g$ on K.

3.3 Theorem: *Let g be upper semicontinuous on Ω; then the following are equivalent:*

(i) *g is subharmonic on Ω.*

(ii) *If D is a closed ball in Ω, f a real harmonic polymonial such that $g \leq f$ on ∂D, then $g \leq f$ on D.*

(iii) *For $\delta > 0$, let $\Omega_\delta = \{z\colon \inf w \notin \Omega |z - w| > \delta\}$, then*

(3.3*) *$g(z) \int_0^\delta d\mu(r) \leq \int_S \int_0^\delta g(z + rw)\, d\mu(r)\, d\omega(w)$ for each positive Borel measure μ on $[0, \delta]$, $z \in \Omega_\delta$.*

(iv) *For each $\delta > 0$, $z \in \Omega_\delta$ there exists a positive Borel measure μ on $[0, \delta]$ such that (3.3*) holds and $\mu(0, \delta] \neq 0$.*

Proof: (i) implies (ii) and (iii) implies (iv) are trivial. Observe that if f is a harmonic polynomial then $\int_S f(z + rw)\, d\omega(w) = f(z)$ for any $z \in \mathbf{C}^n$, $r \geq 0$. Now imitate the proof given in [Hö, p. 17, 1.6.3] replacing "trigonometric polynomials" by "harmonic polynomials," and

$$\frac{1}{2\pi}\int_0^{2\pi} f(z + re^{i\theta})\, d\theta$$

by $\int_S f(z + rw)\, d\omega(w)$, and making other similar changes. □

3.4 Theorem: *Let g be subharmonic on Ω and not identically $-\infty$ in any component of Ω then $g \in L^1_{loc}(\Omega)$ (that is, $\mathbf{R}^{2n}$-Lebesgue integrable on each compact subset of Ω), and thus $g \not\equiv -\infty$* a.e.

Proof: The proof is as in [Hö, 1.6.9, p. 19] where we use $d\mu(r) = r^{2n-1}\, dr$ in (3.3*), since $\mathbf{R}^{2n}$-Lebesgue measure is a constant multiple of $r^{2n-1}\, dr\, d\omega$. □

3.5: Recall the definition of the Poisson kernel for $\mathbf{R}^{2n}$ [9.6.11],

$$P[f](z) = \int_S f(w)\,(1 - |z|^2)\,(1 - 2\mathrm{Re}\langle z, w\rangle + |z|^2)^{-n}\, d\omega(w).$$

Denote $(1 - |z|^2)(1 - 2\mathrm{Re}\langle z, w\rangle + |z|^2)^{-n}$ by $P(z, w)$.

Let g be subharmonic on B_n, $g \not\equiv -\infty$, $0 \leq r < 1$ and define $h_r(z) = \int_S P(z/r, w) g(rw)\, d\omega(w)$ for $|z| < r$, then h_r is harmonic in rB_n and is the least harmonic majorant of g in rB_n, by a standard argument.

3.6 Theorem: *Let g be subharmonic on B_n, $g \not\equiv -\infty$, and* $\sup_{0 \leq r < 1} \int_S g^+(rz)\, d\omega(z) < \infty$ *(where $g^+(z) =$* $\max(0, g(z))$*), then*

(i) *$h_r \geq g$ in rB_n*

(ii) *if $r < s$, then $h_r \leq h_s$ in rB_n, and $\int_S g(rz)\, d\omega(z) \leq \int_S g(sz)\, d\omega(z)$*

(iii) $\lim_{r \to 1} h_r(z) = h(z)$ *exists for each $z \in B_n$ and h is the least harmonic majorant of g in B_n.*

(iv) $\lim_{r \to 1} h(rz) = h^*(z)$ *exists for ω-almost all $z \in S$, $h^* \in L^1(S)$ and there exists a real $\mu_s \in M(S)$ such that $\mu_s \perp \omega$ and $h = P[h^* + d\mu_s]$.*

(v) *if $\lim_{r \to 1} g(rz) = g^*(z)$ exists* a.e. *on S then $g^* = h^*$* a.e. *on S.*

(vi) *$g(rz)\, d\omega(z)$ and $h(rz)\, d\omega(z)$ both converge weak-* (in the dual of $C(S)$) to $\mu = h^* \cdot \omega + \mu_s$ as $r \to 1$ and μ is called the boundary measure of g.*

Proof: We use the radial convergence theorem for Poisson integrals (9.6.14) and suitably modify the proof given in [RFT, p. 41-3], replacing U^n, $\mathbf{T}^n$ by B_n, S respectively. □

3.7 Proposition: *Under the above hypotheses and notation, the least harmonic majorant of* $|h|$ *is* $P[d|\mu|]$.

Proof: We have $h(z) = \int_S P(z, w)\, d\mu(w)$, $z \in B_n$ so

$$|h(z)| \le \int_S P(z, w)\, d|\mu|(w),$$

thus $P[d|\mu|]$ is a harmonic majorant of $|h|$. But $h(rz)\, d\omega(z)$ converges weak-$*$ to μ as $r \to 1$ so for $f \in C(S)$

$$\begin{aligned}\left|\int_S f\, d\mu\right| &= \lim_{r\to 1}\left|\int_S h(rz) f(z)\, d\omega(z)\right| \\ &\le \lim_{r\to 1}\int_S |h(rz)|\,|f(z)|\, d\omega(z) = \int_S |f|\, dv,\end{aligned}$$

where v is the boundary measure of $|h|$, thus $|\mu| \le v$ and already $|\mu| \ge v$ since $|h| \le P[d|\mu|]$. □

3.8 Proposition: *Under the same hypotheses as* 3.6,

$$\int_S |g(rz) - h^*(z)|\, d\omega(z) \to \|\mu_s\| \text{ as } r \to 1.$$

Proof: Let $f = P[h^*]$, then the boundary measure of $g - f$ is μ_s so the boundary measure of $|g - f|$ is $|\mu_s|$, and the weak-$*$ convergence shows $\int_S |g(rz) - f(rz)|\, d\omega(z) \to \|\mu_s\|$ as $r \to 1$. But $\int_S |f(rz) - h^*(z)|\, d\omega(z) \to 0$ as $r \to 1$ by 9.6.13. The two limits together give the result. □

3.9 Definition: A real convex function ϕ on $\mathbf{R}$ (see B.1.2) is said to be **strongly convex** if $\phi \ge 0$, $\phi(t)/t \to \infty$ as $t \to \infty$, and $\phi(-\infty) = \lim_{t\to-\infty}\phi(t)$.

If g is subharmonic then by Jensen's inequality, $\phi \circ g$ is subharmonic.

3.10 Theorem: *Let g be subharmonic on B_n, ϕ be a strongly convex function, and suppose that $C = \sup_{0\le r<1}\int_S \phi \circ g(rz)\, d\omega(z) < \infty$, then g has a boundary measure $\mu = h^* \cdot \omega + \mu_s$ with $\mu_s \le 0$, and the boundary measure of $\phi \circ g$ is absolutely continuous and equals $(\phi \circ h^*) \cdot \omega$.*

Proof: Now $\phi(t) \ge t$ for sufficiently large t so

$$\sup_{0\le r<1} \int_S g^+(rz)\, d\omega(z) < \infty,$$

so g has a boundary measure $h^* \cdot \omega + \mu_s$. Let E be open in S and let $f \in C(S)$ with $0 \le f \le 1$ and $f = 0$ off E. Put

$$v(s) = \sup_{s\le t} \frac{t}{\phi(t)}$$

for $s > 0$, then $v(s) \to 0$ as $s \to \infty$. Then

$$\int_S g(rz) f(z)\, d\omega(z) \leq v(s) \int_S \phi \circ g(rz) f(z)\, d\omega(z) + s \int_S f(z)\, d\omega(z),$$

since $t \leq v(s)\phi(t) + s$ for all $t \in \mathbf{R}$; and by hypothesis, letting $r \to 1$, $\int_S f(z)\, d\mu(z) \leq v(s)C + s\omega(E)$. Now let $s = (\omega(E))^{-1/2}$, then the right-hand side tends to 0 as $\omega(E) \to 0$, thus $\mu_s \leq 0$.

Let ν be the boundary measure of $\phi \circ g$ then

$$\int_S \phi \circ g(rz) f(z)\, d\omega(z) \to \int_S f(z)\, d\nu(z) \text{ as } r \to 1 \text{ for all } f \in C(S).$$

But there exists a sequence $r_m \xrightarrow{m} 1$ such that $g(r_m z) \xrightarrow{m} h^*(z)$ a.e., since $h(0) - h_r(0) = \int_S (h(rz) - g(rz))\, d\omega(z) \to 0$ and $h(rz) \to h^*(z)$ a.e. on S, as $r \to 1$. Now take $f \geq 0$ and apply Fatou's lemma to get

$$\int_S f(z)\, \phi \circ h^*(z)\, d\omega(z) \leq \int_S f\, d\nu,$$

thus $(\phi \circ h^*) \cdot \omega \leq \nu$ and $\phi \circ h^* \in L^1(S)$. Further, $g \leq P[h^* + d\mu_s] \leq P[h^*]$ so by Jensen's inequality

$$\phi \circ g(z) \leq \int_S P(z, w)\, \phi \circ h^*(z)\, d\omega(z) = P[\phi \circ h^*]$$

for $z \in B_n$. But v is the boundary measure of $\phi \circ g$, so this says $v \leq (\phi \circ h^*) \cdot \omega$, and thus $\nu = (\phi \circ h^*) \cdot \omega$. □

3.11: The proofs of Propositions 3.7 and 3.8 are taken from Gårding and Hörmander [1], and Theorem 3.10 is due to them.

4. H^p-Theory

These theorems are the B_n version of the theorems in [RFT, p. 44-53].

4.1 Lemma: *Let $f \in L^1(S)$ or let f be measurable and nonnegative on S, then*

$$\int_S d\omega(z) \frac{1}{2\pi} \int_{-\pi}^{\pi} f(e^{i\theta} z)\, d\theta = \int_S f\, d\omega.$$

Proof: $\int_S f(e^{i\theta} z)\, d\omega(z)$ is independent of θ by the $\mathbf{U}(n)$-invariance of ω, thus Fubini's theorem proves the lemma. □

4.2 Theorem: *Let $f \in H(B_n)$, $f \not\equiv 0$ then $\log |f|$ is subharmonic on B_n.*

Proof: The function $\log|f|$ is upper semicontinuous. Let

$$\{z: |z - z_0| \leq r\} \subset B_n$$

then by 3.3 (iv) it suffices to show that

$$\log|f(z_0)| \leq \int_S \log|f(z_0 + rz)|\, d\omega(z).$$

The inequality is trivial if $f(z_0) = 0$. If $f(z_0) \neq 0$ then for each $z \in S$ let $g_z(\lambda) = f(z_0 + r\lambda z)$. Then g_w is analytic on a neighborhood of $\bar{U}$ in **C** so Jensen's inequality [RRC, p. 300] shows that

$$\log|f(z_0)| = \log|g_z(0)| \leq \frac{1}{2\pi}\int_{-\pi}^{\pi} \log|g_z(e^{i\theta})|\, d\theta$$

$$= \frac{1}{2\pi}\int_{-\pi}^{\pi} \log|f(z_0 + re^{i\theta}z)|\, d\theta$$

Now integrate over $z \in S$ and use Lemma 4.1. □

4.3 Definition: Let $N(B_n)$ (the **Nevanlinna class**) be the set of $f \in H(B_n)$ such that

$$\sup_{0 \leq r < 1} \int_S \log^+ |f(rz)|\, d\omega(z) < \infty.$$

Let $N_*(B_n)$ consist of those $f \in H(B_n)$ for which there is a strongly convex function ϕ such that

$$\sup_{0 \leq r < 1} \int_S \phi(\log|f(rz)|)\, d\omega(z) < \infty.$$

4.4 Theorem: *Let $f \in N(B_n)$, then for almost all $z \in S$, the slice function $f_z \in N(U)$ (where $f_z(\lambda) = f(\lambda z)$, $\lambda \in U$), and $\lim_{r \to 1} f(rz) = f^*(z)$ exists* a.e. *on S.*

Proof: By 4.1, $f_z \in N(U)$ for almost all $z \in S$ and so $\lim_{r \to 1} f_z(re^{i\theta})$ exists for almost all θ [RFT, p. 45, 3.3.3]. □

4.5 Theorem: *Let $f \in N(B_n)$, $f \neq 0$, then f^* exists* a.e. *on S, $\log|f^*| \in L^1(S)$, and there is a real singular ($\perp\omega$) measure μ_s on S such that the least harmonic majorant of $\log|f|$, denoted by $u[f]$, equals $P[\log|f^*| + d\mu_s]$ and the following are equivalent:*

(i) $f \in N_*(B_n)$

(ii) $u[f] \leq P[\log|f^*|]$, *that is,* $\mu_s \leq 0$

(iii) $\log|f| \leq P[\log|f^*|]$.

Proof: Theorem 3.6 applied to $\log|f|$ and the existence of f^* gives the first part. If $f \in N_*(B_n)$ then 3.10 gives (ii). Trivially (ii) implies (iii). The theorem [RFT, 3.1.2, p. 37] shows that there exists a strongly convex ϕ such that $\phi(\log|f^*|) \in L^1(S)$. If (iii) holds then $\phi(\log|f(z)|) \leq P[\phi(\log|f^*|)](z)$ for $|z| < 1$, so

$$\int_S \phi(\log|f(rz)|)\, d\omega(z) \leq \int_S \phi(\log|f^*(z)|)\, d\omega(z)$$

by 9.6.13, for $0 \leq r < 1$, which is the condition for $f \in N_*(B_n)$. See also [RFT, 3.3.5, p. 47]. □

4.6 Definition: Let ϕ be a strongly convex function then $H_\phi(B_n)$ is the set of $f \in H(B_n)$ such that $\sup_{0\leq r<1} \int_S \phi(\log|f(rz)|)\, d\omega(z) < \infty$ (then $H_\phi(B_n) \subset N_*(B_n)$). For $\phi(t) = e^{pt}$, $p > 0$, $H_\phi(B_n) = H^p(B_n)$ and

$$\|f\|_p = \sup_{0\leq r<1} \left\{ \int_S |f(rz)|^p\, d\omega(z) \right\}^{1/p}.$$

Let $H^\infty(B_n)$ be the set of bounded analytic functions on B_n with $\|f\|_\infty = \sup_{B_n}|f|$. From the orthogonality relations 2.9 we see that if $f \in H(B_n)$ with the power series

$$\sum_{\alpha\in Z_+^n} a_\alpha z^\alpha$$

then $f \in H^2(B_n)$ if and only if

$$\|f\|_2 = \left\{ \sum_{\alpha\in Z_+^n} \frac{\alpha!(n-1)!}{(n+|\alpha|-1)!} |a_\alpha|^2 \right\}^{1/2} < \infty.$$

4.7 Theorem: *Let $f \in N^*(B_n)$ and let ϕ be strongly convex. Then $f \in H_\phi(B_n)$ if and only if $\phi(\log|f^*|) \in L^1(S)$, and in that case*

$$\lim_{r\to 1} \int_S \phi(\log|f(rz)|)\, d\omega(z) = \int_S \phi(\log|f^*|)\, d\omega.$$

Proof: If $f \in H_\phi(B_n)$ then $\phi(\log|f^*|) \in L^1(S)$ by 3.10. If $\phi(\log|f^*|) \in L^1(S)$ apply Fatou's Lemma; see [RFT, 3.4.2, p. 51]. □

4.8 Theorem: *If $0 < p < \infty$ and $f \in H^p(B_n)$ then*

$$\lim_{r\to 1} \int_S |f(rz) - f^*(z)|^p\, d\omega(z) = 0.$$

Proof: Apply Egoroff's theorem; see [RFT, 3.4.3, p. 51]. □

4.9 Theorem: *If $0 < p < \infty$ and $f \in H^p(B_n)$ then $|f|^p$ has the least harmonic majorant $P[|f^*|^p]$ and the boundary measure of $|f|^p$ is absolutely continuous.*

Proof: Apply 3.10 to $\log|f|$, with $\phi(t) = e^{pt}$. □

4.10 Corollary: *If $\mu \in M(S)$ and $P[d\mu] \in H(B_n)$ then μ is absolutely continuous.*

Proof: Let $f = P[d\mu]$, then $f \in H^1(B_n)$. Then $|f|$ has the boundary measure $|\mu|$ by 3.7, which is absolutely continuous by 4.9. Thus μ is absolutely continuous. □

Another proof would be 4.8 combined with 3.8. The statement of the corollary can be rephrased: If $\mu \in M(S)$, $\phi_{(m-j,-j)} * \mu = 0$ for $m \geq 1$, $1 \leq j \leq m$, then $\mu \ll \omega$ (an F. and M. Riesz theorem).

4.11 Theorem: *Let $1 \leq p \leq \infty$, $f \in L^p(S)$ such that $\phi_{(m-j,-j)} * f = 0$ for $m \geq 1$, $1 \leq j \leq m$, then $g = P[f] \in H^p(B_n)$, $g^* = f$, and $g(z) = \int_S (1 - \langle z, w \rangle)^{-n} f(w)\, d\omega(w)$, for all $z \in B_n$.*

Proof: The hypotheses on f show that $P[f]$ is a uniformly convergent sum of polynomials (in z) for $|z| \leq r$, each $r < 1$, and by 9.6.13, $\int_S |g(rz)|^p\, d\omega(z) \leq \|f\|_p^p$ for $0 \leq r < 1$, for $p < \infty$; and similarly for $p = \infty$. The Cauchy formula holds since both g and the integral have the same Fourier series on $\mathbf{U}(n)$ (for fixed $|z|$). □

4.12 Corollary: *$H^p(B_n)$ is identified with a closed subspace of $L^p(S)$ under the map $f \mapsto f^*$.*

Proof: If $f \in H^p(B_n)$ then $f^* \in L^p(S)$ and $f = P[f^*]$, and $\|f^*\|_p = \|f\|_p$ (see 9.6.13). An L^p-limit argument shows that $\phi_{(m-j,-j)} * f^*(z) = \lim_{r \to 1} \phi_{(m-j,-j)} * f_r(z)$, for all $z \in S$ where $f_r(z) = f(rz)$. The condition

$$\phi_{(m-j,-j)} * f^* = 0, \qquad m \geq 1, \qquad 1 \leq j \leq m$$

shows $H^p(B_n)$ is closed in $L^p(S)$. □

4.13 Example: There exists $f \in H^\infty(B_n)$ which cannot be continued across any part of S. Let $\{w_m\}$ be a sequence in S containing each point of some countable dense set E infinitely often. Let $\{r_m\}$ be a sequence such that $0 < r_m < 1$ and $\sum_{m=1}^{\infty} m(1 - r_m) < \infty$. Define

$$f(z) = \prod_{m=1}^{\infty} \left(\frac{r_m - \langle z, w_m \rangle}{1 - r_m \langle z, w_m \rangle} \right)^m .$$

This converges to a bounded analytic function since

$$\lambda \mapsto \prod_{m=1}^{\infty} \left(\frac{r_m - \lambda}{1 - r_m \lambda} \right)^m$$

is a convergent Blaschke product in U. Further f has a zero of order at least m at $z = r_m w_m$. Suppose f could be continued across a part of S, then there exists $w \in E$ such that f is analytic in a neighborhood of w. But a subsequence $r_{m_j} w_{m_j} \xrightarrow{j} w$ so f and all of its derivatives equals 0 at w, a contradiction (see [Hö, 2.5.5, p. 39]).

4.14 A remark on A(B$_n$): If $f \in A(B_n)$, $n \geq 2$, and $f(z) \neq c$ for all $z \in S$, then $f(z) \neq c$ for all $z \in B_n$. There is a shell $r < |z| < 1$ where $f(z) \neq c$ so $(f(z)-c)^{-1}$ is analytic there. By Hartog's theorem [Hö, 2.3.2, p. 30] there exists $g \in H(B_n)$ such that $g(z) = (f(z)-c)^{-1}$ on $r < |z| < 1$. Thus $g(z)(f(z)-c) = 1$ on an open set in B_n, hence everywhere. Further $(f(z)-c)^{-1}$ is continuous on $r < |z| \leq 1$, thus $g \in A(B_n)$. In particular if f is inner and $f \in A(B_n)$, that is, $|f(z)| = 1$ for all $z \in S$, then $1/f(z)$ is also inner, so f is constant (this was pointed out to the authors by W. Rudin). The algebra $A(B_n)$, $n \geq 2$, is an example of a function algebra for which the Shilov boundary S is a proper subset of the maximal ideal space $\bar{B}_n$, and for each $f \in A(B_n)$, $f(\bar{B}_n) = f(S)$.

DeLeeuw [1] has studied maximal ideal spaces of algebras generated by polynomials on circled compact sets in $\mathbf{C}^n$.

APPENDIX A

THE HAAR INTEGRAL

Here we present a proof due to Bredon [1] of the existence and uniqueness of a right invariant integral on a locally compact topological group. This proof does not use the axiom of choice.

1.1 Definition: Let G be a locally compact group. It will be convenient to write G multiplicatively. Let L denote the continuous real valued functions on G with compact support, and let L^+ be those which are nonnegative. For $f \in L$ and $x \in G$, let $R(x)f$ denote the right translate of f by x; that is, $R(x)f(y) = f(yx)$, $y \in G$. For $f, g \in L^+$, we write $f \sim g$ if there are $f_1, \ldots, f_n \in L^+$ and $x_1, \ldots, x_n \in G$ such that $f = \sum_{i=1}^n f_i$ and $g = \sum_{i=1}^n R(x_i)(f_i)$.

1.2 Lemma: *Let $f, g \in L^+$. Suppose $g_1, \ldots, g_n \in L^+$ are such that $g = \Sigma g_i$. Then $f \sim g$ if and only if there are $f_i \in L^+ (i = 1, 2, \ldots n)$ such that*

$$f = \sum_{i=1}^{n} f_i \text{ and } f_i \sim g_i.$$

Proof: Let $f \sim g$. Thus $g \sim f$. Hence there are functions $g'_j \in L^+$ and $x_j \in G$ such that $g = \Sigma_j g'_j$ and $f = \Sigma_j R(x_j)(g'_j)$. Let $g_{i,j}$ be defined by

$$g_{i,j}(x) = \begin{cases} \dfrac{g_i(x) g'_j(x)}{g(x)} & \text{if } g(x) > 0 \\ 0 & \text{if } g(x) = 0. \end{cases}$$

Now $g_{i,j}$ is continuous since $g_{i,j}(x) \leq g(x)$. Thus $g_{i,j} \in L^+$ and $g_i = \Sigma_j g_{i,j}$. Let $f_i = \Sigma_j R(x_j)(g_{i,j})$. Then $f_i \in L^+$ and $f_i \sim g_i$. Finally,

$$\sum_i f_i = \sum_i \sum_j R(x_j)(g_{i,j}) = \sum_j \sum_i R(x_j)(g_{i,j})$$
$$= \sum_j R(x_j)(\sum_j g_{i,j}) = \sum_j R(x_j)(g'_j) = f.$$

Now suppose that there are functions $f_i \in L^+$ with $f = \Sigma_i f_i$ and $f_i \sim g_i$. Now there are $f_{i,j} \in L^+$ and $x_{i,j} \in G$ such that $f_i = \Sigma_j f_{i,j}$ and $g_i = \Sigma_j R(x_{i,j})(f_i)$. Thus $f = \Sigma_i f_i = \Sigma_{i,j} f_{i,j}$ and $g = \Sigma_i g_i = \Sigma_{i,j} R(x_{i,j})(f_i)$. Hence $f \sim g$. □

1.3 Lemma: $\sim$ *is an equivalence relation.*

Proof: Symmetry and reflexivity are immediate. For transitivity suppose that $f \sim g$ and $g \sim h$. Thus there are functions g_i and elements y_i with $g = \Sigma_i g_i$ and $h = \Sigma_i R(y_i)(g_i)$. Lemma 2 yields f_i such that $f = \Sigma_i f_i$ where $f_i \sim g_i$. Hence $f_i \sim R(y_i)(g_i)$. Lemma 2 now implies that $f \sim h$ since $h = \Sigma_i R(y_i)(g_i)$. □

1.4 Lemma: Let $f \in L^+$ *and* U *any open subset of* G. *Then there is* $\phi \in L^+$ *with* spt $(\phi) \subset U$ *and* $f \sim \phi$.

Proof: Let V be open with $\overline{V}$ a compact subset of U. Let $\bigcup_{i=1}^{n} Vx_i$ be a finite cover of spt (f). Let $h_i \in L^+$ be such that $h_i(x) = 1$ on Vx_i and such that spt $(h_i) \subset Ux_i$. Let $h = \Sigma_i h_i$. Then $h(x) \geq 1$ for $x \in$ spt (f). Let $f_i = f h_i/h$ on spt (f) and 0 off spt (f). Now $f_i \in L^+$, $f = fh/h = \Sigma f_i$, and spt $(f_i) \subset$ spt (h_i) $\subset Ux_i$. Thus we have written f as Σf_i with spt (f_i) contained in Ux_i. Let $\phi_i = R(x_i^{-1})(f_i)$. Now spt $(\phi_i) \subset U$, $\phi_i \sim f_i$ and by Lemma 2, $\phi = \Sigma \phi_i$ is equivalent to f. □

1.5 Definition: Let $f, g \in L^+$. Write $f \succ g$ provided there exists $f' \sim f$ and $g' \sim g$ such that $f'(x) \geq g'(x)$ for all x.

1.6 Lemma: (1) $f \succ g$ *if and only if* $f = f_1 + f_2 (f_i \in L^+)$ *with* $f_1 \sim g$.

(2) $f \succ g$ *if and only if there is* $f' \sim f$ *with* $f'(x) \geq g(x)$ *for all* x.

(3) $f \succ g$ *and* $g \succ h$ *imply that* $f \succ h$.

(4) *For any* $f, g \in L^+$ *with* $g \not\equiv 0$, *there is a number* s *such that* $sg \succ f$.

(5) $f_i \succ g_i (i = 1, 2)$ *imply that* $f_1 + f_2 \succ g_1 + g_2$.

(6) $f \succ g$ *and* $s > 0$ *imply that* $sf \succ sg$.

Proof: If $f \succ g$, then $f \sim f'$, $g \sim g'$, and $f'(x) \geq g'(x)$. Thus there is $h' \in L^+$ such that $f \sim f' = g' + h'$. Lemma 2 yields f_1, f_2 such that $f = f_1 + f_2$ and

$f_1 \sim g'$ and $f_2 \sim h'$. Conversely, if $f = f_1 + f_2$ with $f_1 \sim g$, then $f \sim g + f_2$ by Lemma 2. Since $f \sim g + f_2$, $g \sim g$, and $(g + f_2)(x) \geq g(x)$, we have that $f \succ g$. Thus (1) holds.

If $f \succ g$, then by (1) there is f_1, f_2 with $f = f_1 + f_2$ and $f_1 \sim g$. Let $f' = g + f_2$. Then $f' \sim f$ and $f'(x) \geq g(x)$. Conversely, if $f' \sim f$ and $f'(x) \geq g(x)$, then $f \succ g$ by the definition. Thus (2) holds.

If $g \succ h$, then write $g = g_1 + g_2$ with $g_1 \sim h$. Now (2) yields f' such that $f' \sim f$ and $f'(x) \geq g(x) \geq g_1(x)$. Thus $f \succ g_1$. Hence $f = f_1 + f_2$ with $f_1 \sim g_1$. Since $g_1 \sim h$, we have that $f \succ h$ by (1). Thus (3) holds.

Let $g \neq 0$. Let $U = \{x \in G : g(x) > 0\}$. By Lemma 4, there is $\phi \in L^+$ with $\operatorname{spt}(\phi) \subset U$ and $f \sim \phi$. Let $t = \min\{g(x) : x \in \operatorname{spt}(\phi)\}$. There is an s such that $st \geq \max\{\phi(x) : x \in G\}$. Hence $sg(x) \geq \phi(x)$. Thus $sg \succ f$. Thus (4) holds.

If $f_1 \succ g_1$ and $f_2 \succ g_2$, then there are f_i', g_i' $(i = 1, 2)$ such that $f_i \sim f_i'$, $g_i \sim g_i'$ and $f_i'(x) \geq g_i'(x)$. Letting $f = f_1 + f_2$ and $g = g_1 + g_2$, we have that $f \sim f_1' + f_2'$, $g \sim g_1' + g_2'$ and $(f_1' + f_2')(x) \geq (g_1' + g_2')(x)$; that is, $f_1 + f_2 \succ g_1 + g_2$. Thus (5) holds.

If $f \succ g$, then there are $f', g' \in L^+$ with $f \sim f'$, $g \sim g'$, and $f'(x) \geq g'(x)$. Now $sf'(x) \geq sg'(x)$, $sf \sim sf'$, and $sg \sim sg'$. Hence $sf \succ sg$. Thus (6) holds. □

1.7 Lemma: (The Marriage Problem, Halmos and Vaughan[1]): *Suppose A and B are finite sets and ϕ is a function defined on 2^A with values in 2^B such that $\phi(E_1 \cup E_2) = \phi(E_1) \cup \phi(E_2)$ for all $E_1, E_2 \subset A$ and* card $\phi(E) \geq$ card *E for all $E \subset A$ (where 2^A, 2^B denote the sets of all subsets of A, B respectively, and* card *denotes cardinality). Then there exists a one-to-one function $f: A \to B$ such that $f(x) \in \phi(\{x\})$ for all $x \in A$.*

Proof: Let card $A = n$. The result is trivial for $n = 1$. Now suppose the result is true for $n = 1, 2, \ldots, m$ for some $m \in Z_+$. Let $n = m + 1$. If card $\phi(E) \geq 1 +$ card E for all $E \subset A$ such that $1 \leq$ card $E \leq m$, then for an arbitrary $x_0 \in A$ define $f(x_0)$ to be some point in $\phi(\{x_0\})$ and apply the induction hypothesis to $A \backslash \{x_0\}$, $B \backslash \{f(x_0)\}$, and the appropriately modified ϕ.

Otherwise there is some subset $E \subset A$, $1 \leq$ card $E \leq m$, such that card $\phi(E) =$ card E. The induction hypothesis gives the existence of the required function f on E, and $f(E) = \phi(E)$. We now claim that $\phi' : 2^{A \backslash E} \to 2^{B \backslash f(E)}$, where $\phi'(F) = \phi(F) \backslash \phi(E)$ for $F \subset A \backslash E$, satisfies the induction hypothesis. For if not, then there exists $F \subset A \backslash E$ such that card $(\phi(F) \backslash \phi(E)) <$ card F and thus card $\phi(E \cup F) =$ card $(\phi(E) \cup \phi(F)) =$ card $\phi(E) +$ card $(\phi(F) \backslash \phi(E)) <$ card $E +$ card $F =$ card $(E \cup F)$ contrary to the hypothesis on ϕ. Now card $(A \backslash E) \leq m$ and the proof is finished by induction. □

1.8 Lemma: *Let C be a compact subset of G and N a compact neighborhood of the identity, e, in G. Let $U \subset N$ be a symmetric open neighborhood of e, and let $\{x_i U : i \in I\}$ (I a finite index set) be a covering of CN by left translates* *of U with a minimal number of elements. For any $x \in G$, let $J = J(x) = \{j \in I : xx_j \in C\}$. Then there exists a one-to-one mapping σ of J into I such that $x_{\sigma(j)} \in xx_jU^2$ for all $j \in J$.*

Proof: Let $j_1, \ldots, j_n$ be distinct elements of J and assume

$$xx_{j_1}U \cup \ldots \cup xx_{j_n}U$$

meets the sets $x_{i_1}U, \ldots, x_{i_m}U$ and no others. Then

$$\bigcup_{k=1}^{n} xx_{j_k}U \subset \bigcup_{k=1}^{m} x_{i_k}U$$

since $xx_jU \subset CU \subset CN$ for $j \in J$. Thus $n \leq m$ for otherwise one could replace the sets $x_{j_k}U$ ($k = 1, \ldots, n$) by the sets $x^{-1}x_{i_k}U$ ($k = 1, \ldots, m$) and obtain a smaller covering of CN. The previous lemma yields a one-to-one mapping $\sigma : J \to I$ such that $xx_jU \cap x_{\sigma(j)}U \neq \emptyset$.

Since U is symmetric, $x_{\sigma(j)} \in xx_jU^2$. □

1.9 Lemma: *Let K be a compact subset of G and $\varepsilon > 0$. Let $g \in L^+$, $g \neq 0$. There exists $x_i \in G$ such that*

$$\left| \frac{\Sigma g(xx_i)}{\Sigma g(x_i)} - 1 \right| < \varepsilon \quad \left(or \left| \frac{\Sigma g(x_ix)}{\Sigma g(x_i)} - 1 \right| < \varepsilon \right)$$

for all $x \in K$. Moreover, the x_i can be chosen so that the above inequality is simultaneously true for any finite number of given functions $g \in L^+$.

Proof: For any two sets U, V with nonempty interior and compact closure we let $[U, V]$ be the least number of left translates of U needed to cover V. Now $[U, W] \leq [U, V][V, W]$.

Let $C = \text{spt}(g) \cup K^{-1}(\text{spt}(g))$. Let $a = \| g \|_\infty$, and let $V = \{x : g(x) \geq 2a/3\}$ and $W = \{x : g(x) \geq a/3\}$. Let N be a compact neighborhood of e in G such that $VN \subset W$. Choose $\delta > 0$ such that

$$\delta < \frac{a\varepsilon}{3[V, CN]}.$$

Choose a symmetric neighborhood U of e with $U \subset N$ such that $x^{-1}y \in U^2$ implies $|g(x) - g(y)| < \delta$.

Let $x_i \in G$ be chosen such that $\{x_iU\}_{i \in I}$ be a minimal covering of CN. Let $x \in K$ be fixed but arbitrary. Since $C \supset \text{spt}(g)$, $g(xx_i) \neq 0$ implies $xx_i \in C$. Thus $i \in J = J(x)$. Let σ be as in Lemma 8. Thus $x_{\sigma(i)} \in xx_iU^2$ which implies that $|g(x_{\sigma(i)}) - g(xx_i)| < \delta$. Hence

$$\sum_{i \in I} g(xx_i) \leq \sum_{i \in J} [g(x_{\sigma(i)}) + \delta] \leq \sum_{i \in I} [g(x_i) + \delta]$$
$$= \sum_{i \in J} g(x_i) + \delta[U, CN].$$

If $x_iU \cap V \neq \emptyset$, then $x_i \in VU \subset VN \subset W$ and hence

$$\sum_{i \in I} g(x_i) \geq [U, V]\frac{a}{3}.$$

Thus,

$$\frac{\delta[U, CN]}{\Sigma g(x_i)} \leq \frac{3\delta[U, CN]}{a[U, V]} \leq \frac{\delta 3[V, CN]}{a} < \varepsilon.$$

Consequently,

$$\frac{\Sigma g(xx_i)}{\Sigma g(x_i)} < 1 + \varepsilon.$$

Now put $C' = xC = x\,\mathrm{spt}\,(g) \cup xK^{-1}\,\mathrm{spt}\,(g) \supset \mathrm{spt}\,(g)$. The sets xx_iU form a minimal cover of $xCN = C'N$. Put $y_i = xx_i$. Then as above,

$$\begin{aligned}\Sigma g(x^{-1}y_i) &\leq \Sigma g(y_i) + \delta[U, C'N] \\ &= \Sigma g(y_i) + \delta[U, CN].\end{aligned}$$

Thus $\Sigma g(xx_i) \geq \Sigma g(x_i) - \delta[U, CN]$. It now follows that

$$\frac{\Sigma g(xx_i)}{\Sigma g(x_i)} > 1 - \varepsilon.$$

Similarly one can obtain the other sided form.

Finally, we note that the proof would work for $g_1, \ldots, g_n$ by changing C to

$$\bigcup_{i=1}^{n} \mathrm{spt}\,(g_i) \cup K^{-1} \bigcup_{i=1}^{n} \mathrm{spt}\,(g_i)$$

since we need only that $C \supset \mathrm{spt}\,(g_i)$ to make the above proof work. □

1.10 Lemma: *If $f \sim g$ and $\varepsilon > 0$, then there exists $y_j \in G, 1 \leq j \leq n$, such that*

$$\left| \frac{\Sigma g(y_j)}{\Sigma f(y_j)} - 1 \right| < \varepsilon.$$

Proof: Let $f = \Sigma f_i$ and $g(x) = \Sigma f_i(xx_i)$ for all x. By the previous lemma, with $K = \{x_i\}$ and $\{f_i\}$, there exists $y_j \in G$ such that

$$\left| \frac{\Sigma_j f_i(y_j x_i)}{\Sigma_j f_i(y_j)} - 1 \right| < \varepsilon \text{ for all } i.$$

Thus

$$\left| \frac{\Sigma_j g(y_j)}{\Sigma_j f(y_j)} - 1 \right| = \left| \frac{\Sigma_i \Sigma_j f_i(y_j x_i)}{\Sigma_i \Sigma_j f_i(y_j)} - 1 \right| < \varepsilon$$

since for a_i, b_i positive,

$$\left|\frac{a_i}{b_i} - 1\right| < \varepsilon$$

implies that

$$\left| \sum_i a_i - \sum_i b_i \right| \leq \sum_i |a_i - b_i| < \varepsilon \, \Sigma b_i$$

and thus

$$\left|\frac{\Sigma_i a_i}{\Sigma_i b_i} - 1\right| < \varepsilon. \quad \square$$

1.11 Theorem: *If $f \in L^+$, $f \neq 0$, $s > 0$, then $f \succ sf$ implies that $s \leq 1$.*

Proof: Suppose $f \succ sf$. Hence $f \sim g$ with $g(x) \geq sf(x)$ for all x. For $\varepsilon > 0$, there are points y_i such that

$$1 + \varepsilon \geq \frac{\Sigma_i \, g(y_i)}{\Sigma_i f(y_i)} \geq \frac{s\Sigma_i f(y_i)}{\Sigma_i f(y_i)} = s. \quad \square$$

1.12 Corollary: *If $f, g \in L^+$ with $g \neq 0$, then* $\sup\{s \mid sg \prec f\} \leq \inf\{s \mid sg \succ f\}$.

Proof: Let $sg \prec f$ and $f \prec tg$, $0 < s, t$. Then $sg/t \prec g$. By Theorem 1.11, $s/t \leq 1$ so $s \leq t$. $\square$

1.13 Lemma: *Given $f \in L^+$ and $\varepsilon > 0$, there exists a compact symmetric neighborhood, U, of e such that for any $g \in L^+$, $g \neq 0$, with* spt $(g) \subset U$ *we can find constants $c_i \geq 0$ and points $y_i \in G$, $1 \leq i \leq n$, such that*

$$| f(x) - \Sigma c_i g(xy_i)| < \varepsilon$$

for all $x \in G$. Furthermore, spt $(\Sigma c_i R(y_i)\,(g)) \subset U^2$ spt (f).

Proof: Let U be a compact symmetric neighborhood of e such that $xy^{-1} \in U$ implies $| f(x) - f(y)| < \varepsilon'$ (ε' will be chosen later). Let K be a compact subset of G containing U^2 spt (f). Let g have support in U. Lemma 9 shows that there are $y_i \in G$, $1 \leq i \leq n$, such that for $x \in K$

$$\left|\frac{\Sigma g(xy_i)}{\Sigma g(y_i)} - 1\right| < \varepsilon''$$

(ε'' will also be chosen later).

For $x \in K$,

$$\left| f(x) - \frac{\Sigma f(x) g(xy_i)}{\Sigma g(y_i)}\right| \leq f(x)\, \varepsilon''.$$

Now $g(xy_i) \neq 0$ implies that $xy_i \in U$ and thus

$$|f(x) - f(y_i^{-1})| < \varepsilon'.$$

Thus for $x \in G$,

$$\left| \frac{\Sigma f(x)g(xy_i)}{\Sigma g(y_i)} - \frac{\Sigma f(y_i^{-1})g(xy_i)}{\Sigma g(y_i)} \right| \leq \frac{\Sigma |f(x) - f(y_i^{-1})| g(xy_i)}{\Sigma g(y_i)}$$

$$\leq \varepsilon' \frac{\Sigma g(xy_i)}{\Sigma g(y_i)} \leq \varepsilon' (1 + \varepsilon'').$$

Thus for $x \in K \supset \mathrm{spt}\,(f)$,

$$(*) \left| f(x) - \frac{\Sigma f(y_i^{-1})g(xy_i)}{\Sigma g(y_i)} \right| \leq f(x)\,\varepsilon'' + \varepsilon' (1 + \varepsilon'').$$

Choose ε' and ε'' such that $\| f \|_\infty \varepsilon'' + \varepsilon' (1 + \varepsilon'') < \varepsilon$. Let

$$c_i = \frac{f(y_i^{-1})}{\Sigma g(y_i)}.$$

Suppose $x \notin U^2\,\mathrm{spt}\,(f)$ but $x \in \mathrm{spt}\,(R(y_i)\,(g)) \subset Uy_i^{-1}$. Now

$$\mathrm{spt}\,(R(y_i)\,(g)) \subset Uy_i^{-1} \subset UU^{-1}x = U^2x.$$

If $y \in \mathrm{spt}\,(R(y_i)g) \cap \mathrm{spt}\,(f)$, then $y \in U^2x$ and $y \in \mathrm{spt}\,(f)$. But then $x \in U^2\,\mathrm{spt}\,(f)$ which is not so. Thus $\mathrm{spt}\,(R(y_i)g) \cap \mathrm{spt}\,(f) = \emptyset$ and we may discard the function $R(y_i)\,(g)$. Thus

$$\mathrm{spt}\,(\Sigma c_i R(y_i)g) \subset U^2\,\mathrm{spt}\,(f).$$

Now $|f(x) - \Sigma c_i R(y_i)g(x)| < \varepsilon$ for $x \in K$ by $(*)$. The inequality is now trivial for $x \notin K \supset U^2\,\mathrm{spt}\,(f)$. □

1.14 Lemma: *If $f, g \in L^+$, $f(x) \geq g(x)$ for all x, and $f \not\equiv g$, then there is a function $f' \sim f$ such that $f'(x) > g(x)$ for all $x \in \mathrm{spt}\,(g)$.*

Proof: Let $h = f - g \in L^+$. Now h is positive on some relatively compact open set, U. Let $x_i \in G$ be such that

$$\mathrm{spt}\,(g) \subset \bigcup_{i=1}^{n} Ux_i^{-1}.$$

Let

$$f' = g + \frac{\Sigma R(x_i)h}{n}.$$

Clearly $f'(x) > g(x)$ on spt (g). Now $(1/n)\,\Sigma R(x_i)h \sim h$ and so $f' \sim g + h = f$. Thus $f' \sim f$. □

1.15 Theorem: *If $f, g \in L^+$, $g \neq 0$, and $\varepsilon > 0$, then there is a number t such that $f \prec tg \prec (1 + \varepsilon)f$.*

Proof: By Lemma 14, there is a function $h \sim (1 + \varepsilon/2)f$ such that $h(x) > f(x)$ for $x \in \text{spt}(f)$. Now $h \prec (1 + \varepsilon)f$. Thus there is $f' \in L^+$ such that $f' \sim (1 + \varepsilon)f$ with $h(x) \leq f'(x)$. Apply Lemma 14 to h and f' to get k such that $k \sim f' \sim (1 + \varepsilon)f$ and $h(x) < k(x)$ on spt (h). Let C be a compact set with spt $(h) \subset$ int (C) and $C \subset$ int (spt (k)). Let U be a symmetric neighborhood of e such that U^2 spt $(h) \subset C$ as in Lemma 13. Let $\phi \sim g$ be such that spt $(\phi) \subset U$ by Lemma 4.

Let $\delta > 0$ be smaller than $\inf\{k(x) - h(x): x \in C\}$ and $\inf\{h(x) - f(x): x \in \text{spt}(f)\}$. By Lemma 13, there exist constants $c_i \geq 0$ and $y_i \in G$ such that

$$|h(x) - \Sigma c_i \phi(xy_i)| < \delta$$

where spt $(\Sigma c_i R(y_i)\phi) \subset U^2$ spt $(h) \subset C$. Thus $f(x) \leq h(x) - \delta \leq \Sigma c_i \phi(xy_i)$ for $x \in \text{spt}(f)$; and $\Sigma c_i \phi(xy_i) \leq h(x) + \delta \leq k(x)$ for $x \in C$. Hence for all $x \in G$, $f(x) \leq \Sigma c_i \phi(xy_i) \leq k(x)$. Thus $f \prec tg \prec (1 + \varepsilon)f$ where $t = \Sigma c_i$. □

1.16 Corollary: $\sup\{s: sg \prec f\} = \inf\{s \mid sg \succ f\}$.

Proof: Let t be as in Theorem 15. Then $\inf\{s: sg \succ f\} \leq t \leq \sup\{s: sg \prec (1 + \varepsilon)f\} = (1 + \varepsilon)\sup\{s: sg \prec f\}$. Now use Corollary 12. □

1.17 Definition: Let $g \in L^+, g \neq 0$. For $f \in L^+$, let $I_g(f) = \sup\{s: sg \prec f\} = \inf\{s: sg \succ f\}$.

1.18 Theorem: *(Existence of the Haar Integral): I_g is a right invariant integral.*

Proof: We need only to check additivity. If $sg \succ f_1$ and $tg \succ f_2$, then $(s + t)g \succ f_1 + f_2$ and so

$$s + t \geq \inf\{s: sg \succ f_1 + f_2\} = I_g(f_1 + f_2).$$

Thus $I_g(f_1) + I_g(f_2) \geq I_g(f_1 + f_2)$.

If $sg \prec f_1$ and $tg \prec f_2$, then $(s + t)g \prec f_1 + f_2$ and so

$$s + t \leq \sup\{s: sg \prec f_1 + f_2\} = I_g(f_1 + f_2).$$

Thus $I_g(f_1) + I_g(f_2) \leq I_g(f_1 + f_2)$. □

1.19 Theorem: *(Uniqueness of the Haar Integral): If I is any right invariant integral and if $g \in L^+$ with $g \neq 0$, then $I = I(g)I_g$.*

Proof: We may assume that $I(g) = 1$. Now if $f \sim h$, then $I(f) = I(h)$. Also, if $f(x) \leq h(x)$ for all $x \in G$, then $I(f) \leq I(h)$. Thus if $f \prec h$, then $I(f) \leq I(h)$. Now, if $sg \succ f$, then $s = sI(g) = I(sg) \geq I(f)$. Thus $I_g(f) \geq I(f)$. Similarly, $I_g(f) \leq I(f)$. Thus $I_g(f) = I(f)$. □

1.20 Historical Notes: Harmonic analysts are indebted to A. Haar [1] who in 1933 showed the existence of a one-sided invariant integral on locally compact groups with a countable open basis. The general case was first shown by Weil [W].

APPENDIX B

INTEGRATION ALGEBRAS

1 Introduction

1.1: For purposes of reference we present in this appendix the definition and elementary properties of $\mathscr{L}^p(\hat{G})$, $1 < p < \infty$, leading up to the Hausdorff–Young–Kunze theorem (that is, $\mathscr{F} : L^p(G) \to \mathscr{L}^{p'}(\hat{G})$, $1 < p < 2$, $p' = p/(p-1)$). We present also a result of Figà–Talamanca and Rider [1] which implies that $\mathscr{F}L^p(G) = \mathscr{L}^{p'}(\hat{G})$, $1 < p < 2$, if and only if G is finite. We use here the notations and definitions of Chapters 7 and 8. The main ingredient of the proofs is the Riesz–Thorin convexity theorem, which we adapt here to the noncommutative integration theory. In Section 1 we state the appropriate form of the Phragmen–Lindelöf theorem and prove a convexity theorem for bilinear forms. Section 2 contains the definition of an integration algebra, proofs of Hölder and Minkowski-type inequalities and the main convexity theorem. The definition of $\mathscr{L}^p(\hat{G})$ $1 < p < \infty$, is given in Section 3, wherein also is the Kunze theorem. Section 4 contains a result of Figà-Talamanca and Rider on random Fourier series and an application.

1.2 Definition: A set $E \subset \mathbf{R}^n$, $n = 1, 2, \ldots$, is **convex** if $x, y \in E$, $0 \leq t \leq 1$, implies $tx + (1-t)y \in E$. A real function ϕ on a convex set $E \subset \mathbf{R}^n$ is said to be convex if $x, y \in E$, $0 \leq t \leq 1$, implies $\phi(tx + (1-t)y) \leq t\,\phi(x) + (1-t)\,\phi(y)$.

1.3 Lemma: *Let E be convex in $\mathbf{R}^n$ and let $\{\phi_l\}$ be a collection of convex* *functions on E, then the function ϕ defined by $\phi(x) = \sup_l \phi_l(x)$ is convex on E.* (See [DS, p. 520].)

1.4 Three Lines Theorem: *Let f be a continuous bounded complex function defined on $E \times \mathbf{R}^n \subset \mathbf{C}^n$ where E is convex in $\mathbf{R}^n$, that is, $z \in E \times \mathbf{R}^n$ if and only if $x \in E$, $y \in \mathbf{R}^n$, $z_j = x_j + iy_j$ for $j = 1, 2, \dots, n$. Suppose further that f is analytic on each open strip in $E \times \mathbf{R}^n$, that is, for each $x', x'' \in E$, the map $\lambda \mapsto g(\lambda) = f(\lambda x' + (1 - \lambda)x'')$ is analytic on $0 < \mathrm{Re}(\lambda) < 1$, $\lambda \in \mathbf{C}$. For each $x \in E$, let $M(x) = \sup\{|f(x + iy)| : y \in \mathbf{R}^n\}$. Then $\log M$ is a convex function on E.* (See [DS, p. 521])

1.5 Remark: For example if $E = (\text{int } E)^-$ and f is analytic in $(\text{int } E) \times \mathbf{R}^n$ (in the several-variable sense) then f satisfies the hypotheses of Theorem 1.4.

1.6 Lemma: *Let F be a bilinear form on $\mathbf{C}^n \times \mathbf{C}^m$ given by*

$$F(z, w) = \sum_{i,j=1}^{n,m} z_i w_j f_{ij}, \quad f_{ij} \in \mathbf{C}.$$

On $\mathbf{C}^n$ there is a collection of metrics given by

$$\phi_p(z) = \left(\sum_{i=1}^{n} \left| z_i \right|^p \alpha_i \right)^{1/p},$$

$0 < p < \infty$, $\phi_\infty(z) = \max |z_i|$, where the $\{\alpha_i\}$ are fixed positive numbers. Similarly on $\mathbf{C}^m$ there is a collection of metrics

$$\psi_q(w) = \left(\sum_{j=1}^{m} \left| w_j \right|^q \beta_j \right)^{1/q},$$

$0 < q < \infty, \psi_\infty(w) = \max |w_j|, (\beta_j > 0, j = 1, 2, \dots, m)$. Let $a = (a_1, a_2) \in \mathbf{R}_+^2$ and let

$$\begin{aligned} M(a_1, a_2) &= \sup\{|F(z, w)|/(\phi_{1/a_1}(z)\,\psi_{1/a_2}(w)) : z, w \neq 0\} \\ &= \sup\{|F(z, w)| : \phi_{1/a_2}(z) \leq 1, \psi_{1/a_2}(w) \leq 1\}. \end{aligned}$$

Then $\log M$ is convex on $\mathbf{R}_+^2$.

Proof: Let $B_1 = \{x \in \mathbf{R}_+^n : \Sigma_i x_i \alpha_i \leq 1\}$ and $B_2 = \{u \in \mathbf{R}^m : \Sigma u_j \beta_j \leq 1\}$. We have

$$M(a_1, a_2) = \sup\left\{ \left| \sum_{i,j=1}^{n,m} x_i^{a_1} z_i u_j^{a_2} w_j f_{ij} \right| : \right.$$

$$\left. \phi_\infty(z) \leq 1, \psi_\infty(w) \leq 1, x \in B_1, u \in B_2 \right\},$$

(observe $(\sum_{i=1}^{n} | x_i^{a_1} z_i |^{1/a_1} \alpha_i)^{a_1} \leq 1$). Further,

$$M(a_1, a_2) = \sup\left\{\left| \sum_{i,j} x_i^{a_1+ib_1} z_i u_j^{a_2+ib_2} w_j f_{ij} \right| : \phi_\infty(z) \leq 1, \psi_\infty(w) \leq 1, x \in B_1, u \in B_2, b \in \mathbf{R}^2 \right\}.$$

For fixed z, w, x, u satisfying the given restrictions let $g(\lambda) = \Sigma_{i,j} x_i^{\lambda_1} z_i u_j^{\lambda_2} w_j f_{ij}$ where $\lambda_j = a_j + ib_j$, then g is bounded and analytic on each set $0 \leq \mathrm{Re}(\lambda_1)$, $\mathrm{Re}(\lambda_2) \leq L$, $0 < L < \infty$, thus by Theorem 1.4, $\log \sup \{ | g(a + ib) | : b \in \mathbf{R}^2 \}$ is convex for $a \in \mathbf{R}_+^2$. Now $\log M(a_1, a_2)$ is the supremum of $\log \sup_b | g(a + ib) |$ over all appropriate z, w, x, b, u, (since log is an increasing function) and thus by Lemma 1.3 the proof is finished. □

2. Integration Algebras

2.1 Definition: Let $\mathscr{A}$ be a $*$-algebra of operators on a Hilbert space with operator norm denoted by $\| \cdot \|_\infty$; and let Tr be a trace defined on $\mathscr{A}$. Then $\mathscr{A}$ is said to be an **integration algebra** if it satisfies the following:

(i) Each $x \in \mathscr{A}$ has a (unique) polar decomposition $x = u | x |$, where $| x | \in \mathscr{A}_+$ (the set of positive elements of $\mathscr{A}$) and u is a partial isometry in $\mathscr{A}$ with $\ker u = \ker | x |$.

(ii) If $x \in \mathscr{A}_+$, $x \neq 0$ then $x = \sum_{i=1}^{n} c_i p_i$ where each $c_i > 0$, p_i is a projection in $\mathscr{A}_+$, $p_i p_j = p_j p_i = 0$ for $i \neq j$, for some $n = 1, 2, \ldots$.

(iii) The trace Tr is finite and faithful, that is, $x \in \mathscr{A}$ implies $\mathrm{Tr}\, x < \infty$, and $x \in \mathscr{A}_+$, $\mathrm{Tr}\, x = 0$ implies $x = 0$.

(iv) $\mathrm{Tr}(xy) = \mathrm{Tr}(yx)$ for all $x, y \in \mathscr{A}$.

(v) $| \mathrm{Tr}(xy) | \leq \| x \|_\infty \mathrm{Tr}(| y |)$ for all $x, y \in \mathscr{A}$.

2.2 Remark: The examples of integration algebras which are discussed here are the space of simple integrable functions on a positive measure space (X, μ) with $\mathrm{Tr}\, f = \int_X f\, d\mu$ (equality meaning μ-almost everywhere) and the space $\mathscr{C}_F(\hat{G})$ of trigonometric polynomials on a compact group G; with trace as defined in 8.3.8. Another example is the algebra of all operators of finite rank on a Hilbert space, with the usual trace.

2.3 Definition: For $0 < p < \infty$, $x \in \mathscr{A}$ let $\| x \|_p = (\mathrm{Tr}(| x |^p))^{1/p}$. Note that if $| x | = \sum_{j=1}^{n} c_j p_j$, as in 2.1(ii), then $| x |^p = \sum_{j=1}^{n} c_j^p p_j \in \mathscr{A}$.

2.4 Proposition: *For $0 < p < \infty$, $\| \lambda x \|_p = | \lambda | \, \| x \|_p$. For $\lambda \in \mathbf{C}$, $x \in \mathscr{A}$ and $\| x \|_p = 0$ implies $x = 0$. Further, $\| x^* \|_p = \| x \|_p$.*

Proof: The first sentence is obvious. If $x \in \mathscr{A}$ then x has the polar decomposition $u|x|$ and $|x^*|^p = u|x|^p u^*$, so $\operatorname{Tr}(|x^*|^p) = \operatorname{Tr}(u|x|^p u^*) = \operatorname{Tr}(u^*u|x|^p) = \operatorname{Tr}(|x|^p)$, since $u^*u|x|^p = |x|^p$. □

2.5 Theorem: *Let $x \in \mathscr{A}$ or $x = I$ (the identity operator), let $y, z \in \mathscr{A}$, and let $1 < p < \infty$, $p' = p/(p-1)$, then*

$$|\operatorname{Tr}(xyz)| \leq \|x\|_\infty \|y\|_p \|z\|_{p'}.$$

Proof: The idea is to interpolate between the known (2.1(v)) cases $p = 1$ and $p = \infty$. We write the polar decompositions $y = u|y| = u(\sum_{i=1}^n c_i p_i)$ and $z = v|z| = v(\sum_{j=1}^m d_j q_j)$ where the $\{p_i\}$ and $\{q_j\}$ are each a mutually orthogonal set of projections (as in 2.1(ii)) in $\mathscr{A}_+$. For $\lambda \in \mathbf{C}^n$, $\mu \in \mathbf{C}^m$, let $y_\lambda = \sum_{i=1}^n \lambda_i u p_i$, $z_\mu = \sum_{j=1}^m \mu_j v q_j$; then

$$\|y_\lambda\|_p = \left(\sum_{i=1}^n |\lambda_i|^p (\operatorname{Tr} p_i)\right)^{1/p}$$

for $0 < p < \infty$ and $\|y_\lambda\|_\infty = \max_i |\lambda_i|$. Also

$$\|z_\mu\|_q = \left(\sum_{j=1}^m |\mu_j|^q (\operatorname{Tr} q_j)\right)^{1/q}$$

for $0 < q < \infty$ and $\|z_\mu\|_\infty = \max_j |\mu_j|$. For $a \in \mathbf{R}_+^2$ let

$$M(a) = \sup\{|\operatorname{Tr}(xy_\lambda z_\mu)| : \lambda \in \mathbf{C}^n, \mu \in \mathbf{C}^m, \|y_\lambda\|_{1/a_1} \leq 1, \|z_\mu\|_{1/a_2} \leq 1\}.$$

Observe that $\operatorname{Tr}(xy_\lambda z_\mu) = \sum_{i,j} \lambda_i \mu_j \operatorname{Tr}(xup_i vq_j)$, so by 1.6, $\log M$ is convex on $\mathbf{R}_+^2$. Now by 2.1(v),

$$|\operatorname{Tr}(xy_\lambda z_\mu)| \leq \|xy_\lambda\|_\infty \|z_\mu\|_1 \leq \|x\|_\infty \|y_\lambda\|_\infty \|z_\mu\|_1$$

thus $M(0, 1) \leq \|x\|_\infty$. Also

$$|\operatorname{Tr}(xy_\lambda z_\mu)| = |\operatorname{Tr}(z_\mu x y_\lambda)| \leq \|z_\mu\|_\infty \|x\|_\infty \|y_\lambda\|_1$$

thus $M(1, 0) \leq \|x\|_\infty$. Hence for $1 < p < \infty$, $M(1/p, 1/p') \leq \|x\|_\infty$. □

2.6 Theorem: *For $1 \leq p < \infty$ and $x \in \mathscr{A}$, $\|x\|_p = \sup\{|\operatorname{Tr}(xy)| : \|y\|_{p'} \leq 1, y \in \mathscr{A}\}$ $(1/p + 1/p' = 1)$; and $\|\cdot\|_p$ is a norm.*

Proof: It is enough to show for $x \neq 0$, that $\|x\|_p = \operatorname{Tr}(xy)/\|y\|_{p'}$ for some $y \neq 0$. Write the polar decomposition $x = u|x|$. If $p = 1$, put $y = u^*$, then $\operatorname{Tr}(xy) = \operatorname{Tr}(u|x|u^*) = \operatorname{Tr}(|x|) = \|x\|_1$, and $\|y\|_\infty = 1$. If

$$1 < p < \infty, \quad \text{put } y = |x|^{p-1} u^*,$$

then

$$\operatorname{Tr}(xy) = \operatorname{Tr}(u|x|^p u^*) = \|x\|_p^p \quad \text{and} \quad \|y\|_{p'} = \|y^*\|_{p'} \quad \text{(by 2.4)}$$
$$= \|u|x|^{p-1}\|_{p'} = (\operatorname{Tr}(|x|^{(p-1)p'}))^{1/p'} = \|x\|_p^{p/p'}.$$

For any $x, y \in \mathscr{A}$ we have

$$\begin{aligned}\| x + y \|_p &= \sup \{ | \operatorname{Tr}((x + y)z) | : \| z \|_{p'} \leq 1\} \\ &\leq \sup \{ | \operatorname{Tr}(xz) | : \| z \|_{p'} \leq 1\} + \sup \{ | \operatorname{Tr}(yz) | : \| z \|_{p'} \leq 1\} \\ &= \| x \|_p + \| y \|_p.\end{aligned}$$

This together with 2.4 shows that $\| \cdot \|_p$ is a norm. □

2.7 Corollary: *For* $1 \leq p < \infty$, $x, y \in \mathscr{A}$, $\| xy \|_p \leq \| x \|_\infty \| y \|_p$.

Proof: We have $\| xy \|_p = \sup \{ | \operatorname{Tr}(xyz) | : \| z \|_{p'} \leq 1\} \leq \| x \|_\infty \| y \|_p$ by 2.5. □

2.8 Theorem: *Let* $\mathscr{A}_1, \ldots, \mathscr{A}_n$ *be integration algebras,* $a \in \mathbf{R}^n_+$ *and let* $B(a) = \{x \in \Sigma\bigoplus_{j=1}^n \mathscr{A}_j : \| x_j \|_{1/a_j} \leq 1,\ 1 \leq j \leq n\}$ *(where* $x = (x_1, \ldots, x_n)$, $x_j \in \mathscr{A}_j$*). Suppose that F is a complex analytic function on* $\Sigma\bigoplus\mathscr{A}_j$ *(that is, analytic on each finite dimensional subspace) and let* $M(a) = \sup \{ | F(x) | : x \in B(a)\}$*, then* log M *is a convex function on* I^n*, where* $I = [0, 1]$.

Proof: Let $B_+(1) = \{x \in B(1, 1, \ldots, 1) : x_j \in \mathscr{A}_{j+},\ 1 \leq j \leq n\}$. For $a \in I^n$, $b \in \mathbf{R}^n$, $x \in B(0)$, $y \in B_+(1)$ we have

$$xy^{a+ib} = ((x_j y_j^{a_j + ib_j})_{j=1}^n) \in B(a)$$

since

$$\| x_j y_j^{a_j + ib_j} \|_{1/a_j} \leq \| x_j y_j^{ib_j} \|_\infty \| y_j^{a_j} \|_{1/a_j} \leq 1.$$

Observe that each $y_j = \sum_{m=1}^{n_j} c_{jm} p_{jm}$, $c_{jm} > 0$, $p_{jm} \in \mathscr{A}_{j+}$ (2.1(ii)) and

$$y_j^{a_j + ib_j} = \sum_{m=1}^{n_j} c_{jm}^{a_j + ib_j} p_{jm}.$$

Conversely if $x \in B(a)$, then $x = u | x | = (u_1 | x_1 |, \ldots, u_n | x_n |)$ with $u \in B(0)$ and $(| x_1 |^{1/a_1}, \ldots, | x_n |^{1/a_n}) \in B_+(1)$. Thus, symbolically, $B(a) = B(0)B_+(1)^{a+ib}$, $b \in \mathbf{R}^n$. Now

$$M(a) = \sup \{ | F(xy^{a+ib}) | : x \in B(0), y \in B_+(1), b \in \mathbf{R}^n\}.$$

Fix $x \in B(0)$, $y \in B_+(1)$, then $y_j = \sum_{m=1}^{n_j} c_{jm} p_{jm}$ (decomposition of 2.1(ii)) $1 \leq j \leq n$ and $xy^{a+ib} = ((\Sigma_m c_{jm}^{a_j + ib_j} x_j p_{jm})_{j=1}^n)$ has values in a bounded subset of a fixed finite dimensional subspace of $\Sigma\bigoplus\mathscr{A}_j$ for all $a \in I^n$, $b \in \mathbf{R}^n$. Thus the map $a + ib \mapsto F(xy^{a+ib})$ is analytic and bounded on $a \in I^n$. By Theorem 1.4, $\log \sup_b | F(xy^{a+ib}) |$ is convex on $a \in I^n$, but log is an increasing function, so $\log M(a) = \sup \{\log \sup_b | F(xy^{a+ib}) | : x \in B(0), y \in B^+(1)\}$ and $\log M$ is convex on I^n by 1.3. □

3. The spaces $\mathscr{L}^p(\hat{G})$, $1 < p < \infty$.

3.1 Definition: For $1 < p < \infty$, $\phi \in \mathscr{L}^\infty(\hat{G})$ let

$$\| \phi \|_p = ((\mathrm{Tr}\,(|\phi|^p))^{1/p} = \left(\sum_\alpha n_\alpha \,\mathrm{Tr}\,(|\phi_\alpha|^p) \right)^{1/p}.$$

Let

$$\mathscr{L}^p(\hat{G}) = \{\phi \in \mathscr{L}^\infty(\hat{G}) \colon \| \phi \|_p < \infty\}.$$

3.2 Proposition: *For $1 < p < \infty$, $\mathscr{L}^p(\hat{G})$ is a Banach space with the norm $\| \cdot \|_p$ and is the completion of $(\mathscr{C}_F(\hat{G}), \| \cdot \|_p)$. Also $\mathscr{L}^p(\hat{G}) \subset \mathscr{C}_0(\hat{G})$.*

3.3 Proposition: *If $1 < p < \infty$, $p' = p/(p-1)$, $\phi \in \mathscr{L}^p(\hat{G})$, $\psi \in \mathscr{L}^{p'}(\hat{G})$, then $\phi\psi \in \mathscr{L}^1(\hat{G})$ and $|\mathrm{Tr}\,(\phi\psi)| \le \| \phi\psi \|_1 \le \| \phi \|_p \| \psi \|_{p'}$.*

3.4 Proposition: *If $1 < p < \infty$, $p' = p/(p-1)$ then the dual of $\mathscr{L}^p(\hat{G})$ may be identified with $\mathscr{L}^{p'}(\hat{G})$ under the pairing $(\phi, \psi) \mapsto \mathrm{Tr}\,(\phi\psi)$ (as in 8.3.9).*

3.5 Theorem: (*Hausdorff-Young-Kunze*): *Let $1 < p < 2$, $p' = p/(p-1)$, $f \in L^p(G)$ then $\mathscr{F} f \in \mathscr{L}^{p'}(\hat{G})$ and $\| \mathscr{F} f \|_{p'} \le \| f \|_p$. Conversely if $\phi \in \mathscr{L}^p(\hat{G})$ then there exists a unique $f \in L^{p'}(G)$ such that $\mathscr{F} f = \phi$, and $\| f \|_{p'} \le \| \phi \|_p$.*

Proof: Let $\mathscr{A}$ be the space of simple integrable functions on G and define a complex linear function F on $\mathscr{A} \oplus \mathscr{C}_F(\hat{G})$ by

$$\begin{aligned} F(f, \phi) &= \sum_{\alpha \in \hat{G}} n_\alpha \,\mathrm{Tr}\,(\hat{f}_\alpha \phi_\alpha) \\ &= \int_G f(x^{-1}) \left(\sum_{\alpha \in \hat{G}} n_\alpha \,\mathrm{Tr}\,(T_\alpha(x)\phi_\alpha) \right) dx \end{aligned}$$

(finite sum).
Now

$$|F(f, \phi)| \le \|\hat{f}\|_\infty \| \phi \|_1 \le \| f \|_1 \| \phi \|_1$$

(by 8.3.9 and 8.4.2(i)) and

$$|F(f, \phi)| \le \|\hat{f}\|_2 \| \phi \|_2 = \| f \|_2 \| \phi \|_2.$$

Let $a \in [0, 1]^2$, and let

$$M(a) = \sup \{ |F(f, \phi)| : f \in \mathscr{A}, \phi \in \mathscr{C}_F(\hat{G}), \quad \| f \|_{1/a_1} \le 1, \quad \| \phi \|_{1/a_2} \le 1 \},$$

then by 2.8, $\log M$ is convex on $[0, 1]^2$. The above remarks show $M(1, 1) \le 1$ and $M(1/2, 1/2) \le 1$, thus $M(1/p, 1/p) \le 1$ for $1 < p < 2$. That is for $f \in \mathscr{A}$,

$\phi \in \mathscr{C}_F(\hat{G}), 1 < p < 2, |\Sigma_\alpha n_\alpha \operatorname{Tr}(\hat{f}_\alpha \phi_\alpha)| \leq \| f \|_p \| \phi \|_{p'}$. Thus by 3.4, $\hat{f} \in \mathscr{L}^{p'}(\hat{G})$ and $\|\hat{f}\|_{p'} \leq \| f \|_p$. But $\mathscr{A}$ is dense in $L^p(G)$ so this inequality extends to all of $L^p(G)$. Conversely for $\phi \in \mathscr{L}^p(\hat{G})$ we see that

$$\left(\int_G |\Sigma_\alpha n_\alpha \operatorname{Tr}(T_\alpha(x)\phi_\alpha)|^{p'} dx\right)^{1/p'} \leq \| \phi \|_p$$

and this defines a bounded map (the inverse Fourier transform) from $\mathscr{L}^p(\hat{G})$ into $L^{p'}(G)$. □

3.6 Historical Note: Theorem 3.5 was proved by Kunze [1] for any locally compact unimodular group in the framework of gage spaces (see also Segal [2, 3], Stinespring [1]). Section 2 is essentially based on Kunze's methods with the simplifications possible due to compactness.

4. A Theorem of Littlewood

4.1: We wish to show for G an infinite compact group that the Fourier transform $\mathscr{F}$ never takes $L^p(G)$ $(1 \leq p < 2)$ onto

$$\mathscr{L}^{p'}(\hat{G}) \left(p' = \frac{p}{p-1}\right).$$

To do this we first must prove the nonabelian analogue of a theorem of Littlewood. This extension of Littlewood's theorem is due to Figà-Talamanca and Rider [1].

4.2 Lemma: *Let $G = \mathbf{U}(n)$ and A be an $n \times n$ matrix. For $s = 1, 2, \ldots,$*

$$\int_G |\operatorname{Tr}(Av)|^{2s} dm_G(v) \leq \frac{(s!)^2}{n^s} (\operatorname{Tr}(A^*A))^s.$$

Proof: Since m_G is left and right invariant we may assume that A is diagonal and positive. Let $A = \operatorname{diag}(a_1, a_2, \ldots, a_n)$ and let $v = (v_{ij})_{1 \leq i, j \leq n}$. Then

$$\begin{aligned}
\int_G |\operatorname{Tr}(Av)|^{2s} dm_G(v) &= \int_G \left| \sum_{i=1}^n a_i v_{ii} \right|^{2s} dm_G(v) \\
&= \int_G \left(\sum_{i=1}^n a_i v_{ii} \right)^s \left(\sum_{j=1}^n a_j \bar{v}_{jj} \right)^s dm_G(v) \\
&= \int_G \left(\Sigma \binom{s}{\beta} a^\beta v^\beta \right) \left(\Sigma \binom{s}{\beta'} a^{\beta'} \bar{v}^{\beta'} \right) dm_G(v)
\end{aligned}$$

(where the sum is taken over $\beta, \beta' \in Z^n_+$ with $|\beta| = |\beta'| = s$ and where $a^\beta v^\beta = a_1^{\beta_1} \ldots a_n^{\beta_n} v_{11}^{\beta_1} \ldots v_{nn}^{\beta_n}$ and similarly for $a^{\beta'} \bar{v}^{\beta'}$)

$$= \Sigma \binom{s}{\beta}\binom{s}{\beta'} a^\beta a^{\beta'} \int_G v^\beta \bar{v}^{\beta'} dm_G(v)$$

$$= \sum_{\beta} \binom{s}{\beta}^2 a^{2\beta} \int_G | v |^{2\beta} \, dm_G(v)$$

(since the integrals are zero for $\beta \neq \beta'$ by the translation invariance of m_G)

$$\leq \sum_{\beta} \binom{s}{\beta}^2 a^{2\beta} \left(\int_G | v_{11} |^{2s} \, dm_G(v) \right)^{\beta_1/s}$$

$$\cdots \left(\int_G | v_{nn} |^{2s} \, dm_G(v) \right)^{\beta_n/s}$$

(by Hölder's inequality)

$$= \sum_{\beta} \binom{s}{\beta}^2 a^{2\beta} \frac{s!(n-1)!}{(n+s-1)!}$$

(by 10.2.9)

$$\leq \frac{s!}{n^s} \left(\max_{\beta} \binom{s}{\beta} \right) \sum_{\beta} \binom{s}{\beta} a^{2\beta}$$

$$\leq \frac{(s!)^2}{n^s} (a_1^2 + \cdots + a_n^2)^s = \frac{(s!)^2}{n^s} (\mathrm{Tr}\,(A^*A))^s. \quad \square$$

4.3 Lemma: *Let G be an infinite compact group and define $\Omega = \Pi_{\alpha \in \hat{G}} \mathrm{U}(n_\alpha)$. Let $f \in L^2(\Omega)$ be such that f has the Fourier series*

$$(*) \quad f(\omega) \sim \sum_{\alpha \in \hat{G}} n_\alpha \, \mathrm{Tr}\,(\hat{f}_\alpha \omega_\alpha)$$

(we use α to denote here the element of $\hat{\Omega}$ given by $\omega \mapsto \omega_\alpha$, the projection map). Then

$$\int_\Omega | f(\omega) |^{2s} \, dm_\Omega(\omega) \leq (s!)^2 \left(\int_\Omega | f(\omega) |^2 \, dm_\Omega(\omega) \right)^s.$$

Proof: Let $f(\omega) = \Sigma_{\alpha \in S} n_\alpha \mathrm{Tr}\,(\hat{f}_\alpha \omega_\alpha)$ where $S = \{\alpha_1, \ldots, \alpha_n\}$ is a finite subset of $\hat{G}$. Now we proceed as in the previous lemma to deduce that

$$\int_\Omega | f(\omega) |^{2s} \, dm_\Omega(\omega) = \sum \binom{s}{\beta} \binom{s}{\beta'} (n_\alpha)^\beta (n_\alpha)^{\beta'} \int_\Omega (\mathrm{Tr}\,(\hat{f}_\alpha \omega_\alpha))^\beta \, \overline{(\mathrm{Tr}\,(\hat{f}_\alpha \omega_\alpha))^{\beta'}} \, dm_\Omega(\omega)$$

$$= \sum_{\beta} \binom{s}{\beta}^2 (n_\alpha)^{2\beta} \int_\Omega | \mathrm{Tr}\,(\hat{f}_\alpha \omega_\alpha) |^{2\beta} \, dm_\Omega(\omega)$$

$$= \sum_{\beta} \binom{s}{\beta}^2 (n_\alpha)^\beta \int_\Omega (n_{\alpha_1})^{\beta_1} | \mathrm{Tr}\,(\hat{f}_{\alpha_1} \omega_{\alpha_1}) |^{2\beta_1} \, dm_\Omega(\omega) \ldots$$

$$\int_\Omega (n_{\alpha_n})^{\beta_n} | \mathrm{Tr}\,(\hat{f}_{\alpha_n} \omega_{\alpha_n}) |^{2\beta_n} \, dm_\Omega(\omega)$$

$$\leq \sum_{\beta} \binom{s}{\beta}^2 (n_\alpha)^\beta (\beta_1 !)^2 (\mathrm{Tr}\,(\hat{f}_{\alpha_1}{}^* \hat{f}_{\alpha_1}))^{\beta_1} \cdots (\beta_n !)^2 (\mathrm{Tr}\,(\hat{f}_{\alpha_n}^* \hat{f}_{\alpha_n}))^{\beta_n}$$

(by the previous lemma)

$$\leq (s!) \Big(\max_\beta \beta! \Big) \sum_\beta \binom{s}{\beta} (n_\alpha \, \mathrm{Tr}\,(\hat{f}_\alpha^* \hat{f}_\alpha))^\beta$$

$$\leq (s!)^2 \Big(\sum_{\alpha \in S} n_\alpha \, \mathrm{Tr}\,(\hat{f}_\alpha^* \hat{f}_\alpha) \Big)^s$$

$$= (s!)^2 \Big(\int_\Omega | f(\omega) |^2 \, dm_\Omega(\omega) \Big)^s$$

(by Theorem 8.4.3).

Now suppose that $f \in L^2(\Omega)$ as in (∗). Let $\{\alpha_1, \alpha_2, \ldots\}$ be the countable subset of $\hat{G} \subset \hat{\Omega}$ which supports $\hat{f}$. Define

$$f_n(\omega) = \sum_{\alpha \in S_n} n_\alpha \, \mathrm{Tr}\,(\hat{f}_\alpha \omega_\alpha)$$

where $S_n = \{\alpha_1, \ldots, \alpha_n\}$. Now $f_n \xrightarrow{n} f$ in $L^2(\Omega)$ and so there is a subsequence $\{n_k\}$ such that $f_{n_k} \xrightarrow{k} f$ pointwise. The result now follows from a simple application of Fatou's lemma. □

4.4 Theorem: *Let G be an infinite compact group. Let $f \in L^2(G)$ and $p < \infty$. For $\omega \in \Omega = \Pi_{\alpha \in \hat{G}} \mathbf{U}(n_\alpha)$ define $\hat{\omega} f \in L^2(G)$ by $(\hat{\omega} f)(x) \sim \Sigma_{\alpha \in \hat{G}} n_\alpha \, \mathrm{Tr}\,(\omega_\alpha \hat{f}_\alpha T_\alpha(x))$ (see 8.4.17). Then $\hat{\omega} f \in L^p(G)$ almost surely (with respect to the probability measure m_Ω).*

Proof: We first let s be an integer such that $s > p/2$. Now consider $(\hat{\omega} f)(x)$ as a function in $L^2(\Omega \times G)$. For fixed $x \in G$ we have by the previous lemma that

$$\int_\Omega | (\hat{\omega} f)(x) |^{2s} \, dm_\Omega(\omega) \leq (s!)^2 \Big(\int_\Omega | (\hat{\omega} f)(x) |^2 \, dm_\Omega(\omega) \Big)^s$$

$$= (s!)^2 \Big(\sum_{\alpha \in \hat{G}} n_\alpha \, \mathrm{Tr}\,(\hat{f}_\alpha^* \hat{f}_\alpha) \Big)^s$$

$$= (s!)^2 \Big(\int_G | f(x) |^2 \, dm_G(x) \Big)^s.$$

Thus

$$\int_\Omega \int_G | (\hat{\omega} f)(x) |^{2s} \, dm_G(x) \, dm_\Omega(\omega) = \int_G \int_\Omega | (\hat{\omega} f)(x) |^{2s} \, dm_\Omega(\omega) \, dm_G(x)$$

$$\leq (s!)^2 \Big(\int_G | f(x) |^2 \, dm_G(x) \Big)^s < \infty.$$

Therefore $\hat{\omega} f \in L^{2s}(G)$ almost surely. □

4.5 Theorem: *Let G be an infinite compact group. Then* $\mathscr{F}(L^p(G))$ *is a proper subset of* $\mathscr{L}^{p'}(\hat{G})$ $(1 < p < 2$ *and* $p' = p/(p-1))$.

Proof: If $\mathscr{F}: L^p(G) \rightarrow \mathscr{L}^{p'}(\hat{G})$ is onto, then the adjoint, which is $\mathscr{F}^{-1}$, takes $\mathscr{L}^p(\hat{G})$ one-to-one and onto $L^{p'}(G)$. Choose $\phi \in \mathscr{L}^2(\hat{G}) \backslash \mathscr{L}^p(\hat{G})$ (here $p < 2$). Let $g = \mathscr{F}^{-1}\phi \in L^2(G)$. Now the previous theorem yields the existence of a unitary operator $\omega \in \Omega \subset \mathscr{L}^\infty(\hat{G})$ such that $f = \check{\omega} g \in L^{p'}(G)$. Now $f \notin \mathscr{L}^p(\hat{G})$ since $\| \hat{f} \|_p = \| \omega \hat{g} \|_p = \| \hat{g} \|_p = \| \phi \|_p = \infty$. □

4.6 Theorem: *Let G be an infinite compact group. Then* $\mathscr{F} L^1(G)$ *is a proper subset of* $\mathscr{C}_0(\hat{G})$.

Proof: Observe that $f \in L^1(G)$ is in the center of $L^1(G)$ if and only if f has the Fourier series $\Sigma_{\alpha \in \hat{G}} n_\alpha c_\alpha \chi_\alpha$, that is, $\hat{f}_\alpha = c_\alpha I_{n_\alpha}$. If $\mathscr{F} L^1(G) = \mathscr{C}_0(\hat{G})$, then for each $\psi \in C_0(\hat{G})$ (functions vanishing at infinity on the discrete space $\hat{G}$) there exists an $f \in L^1(G)$ such that $\hat{f}_\alpha = \psi(\alpha) I_{n_\alpha}$. Hence $C_0(\hat{G})$ is isomorphic to a closed subspace of $L^1(G)$ and is thus weakly sequentially complete; so $\hat{G}$ is finite (as in the proof of 3.2.1). □

APPENDIX C

REMARKS ON RECENT WORK

Kessler [1] has proved a general theorem on semi-idempotent measures on compact abelian groups. Let G be a compact abelian group with a totally ordered dual, then we say $\mu \in M(G)$ is **semi-idempotent** if $\hat{\mu}(\gamma) = 1$ or 0 for all $\gamma > 0$. The theorem says that for any semi-idempotent measure μ there exists an idempotent $\lambda \in M(G)$ such that $\hat{\lambda}(\gamma) = \hat{\mu}(\gamma)$ for all $\gamma > 0$. The proof uses some techniques of Itô and Amemiya (see Chapter 6, Sections 6–8). Helson [2] originally proved this theorem for the case $G = \mathbf{T}$.

Figà–Talamanca and Rider have investigated **lacunary series** questions for compact (nonabelian) groups. As well as the work discussed in Appendix B, they have also considered in [2] series with random coefficients, namely if $\psi \in \mathscr{L}^\infty(\hat{G})$ has the property that $\omega\psi \in \mathscr{F}(L^1(G))$ for ω in a set of positive measure in Ω (defined in B.4.3) then $\psi \in \mathscr{L}^2(\hat{G})$. This paper uses some results of Edwards and Hewitt [1] on the convergence of Fourier series on compact groups. Some earlier work in this area was done by Helgason [2].

Mayer also studied the convergence of Fourier series on compact groups, in particular he showed in [1] that the class of functions on $\mathbf{SU}(2)$ whose Fourier series converge uniformly and absolutely is larger than $A(\mathbf{SU}(2))$. He considered questions of localization of series on $\mathbf{SU}(2)$ in [2, 3, 4].

M. Taylor [1] has proved a Soboleff-type theorem for compact Lie groups, namely, if G is of dimension n, and f has all (distribution sense) derivatives of order $\leq s$ in $L^2(G)$, where s is an even integer $\geq n/2$, then the Fourier series of f converges uniformly and absolutely.

Rider has some results on **central idempotent measures** on nonabelian groups. A measure $\mu \in M(G)$ is said to be central if $\mu(Ex) = \mu(xE)$ for all $x \in G$, E Borel $\subset G$, or equivalently, if $\mu * \nu = \nu * \mu$ for all $\nu \in M(G)$. If G is compact and μ is a central idempotent then $\mu \sim \Sigma_{\alpha\in\hat{G}} n_\alpha s_\alpha \chi_\alpha$ where $s_\alpha = 0$ or 1 for all α (notation of Chapter 7). In [3] Rider describes the sets $S \subset \hat{G}$ such that $S = \{\alpha \in \hat{G} : s_\alpha = 1\}$ for a central idempotent, for a class of groups including the unitary groups. The description is in terms of hypercosets, and the theorem depends on the Cohen theorem (Chapter 6) for abelian groups. In [4] Rider studies central idempotent measures on noncompact groups, and shows that the support group of a central idempotent measure is compact if each neighborhood of e (the identity in the group) contains a neighborhood of e invariant under all inner automorphisms. Recently Rider [5] has shown that only real-analytic functions operate in $A(G)$ (see 7.6) for any infinite compact group (it does not seem to be known whether an infinite compact group must contain an infinite abelian subgroup).

Possible areas for research in compact groups are, of course, suggested by known theorems for compact abelian groups. Along this line the authors have investigated $M_0(G)$ (the space of *measures whose Fourier-Stieltjes transforms vanish at infinity*), Helson sets and Sidon sets.

We define $M_0(G)$ for a compact group G as the set $\{\mu \in M(G) : \hat{\mu} \in \mathscr{C}_0(\hat{G})\}$. In [1] we characterized $M_0(G)$ as the set of measures for which the map $x \mapsto L(x)\mu$ is $\|\cdot\|_\infty$-continuous from G to $M(G)$. We also proved in that paper that $M_0(G)$ is a band (that is, if $\mu \in M_0(G)$ and $\nu \ll \mu$, then $\nu \in M_0(G)$). These results also have a locally compact group setting. We continued our investigation of $M_0(G)$ in [2] where we proved that an open continuous homomorphism from G to H (another compact group) induces a homomorphism of $M_0(G)$ into $M_0(H)$. In [3] we showed that every $\mu \in M_0(G)$ is continuous, and thus no nonzero discrete measure is in $M_0(G)$ (for G infinite).

A **Helson set** P is defined to be a compact set in G such that every continuous function on P extends to a function in the Fourier algebra $A(G)$. In [3], we showed that a Helson set cannot support a nonzero measure in $M_0(G)$ (for G infinite).

We studied Sidon sets for compact groups in [4]. The results in 5.5 for peak sets were shown to hold for compact groups. In that paper we also proved that $A(G)$ is weakly sequentially complete.

BIBLIOGRAPHY

Reference Books

[AS] Abramowitz, M. and Stegun, I. A. *Handbook of Mathematical Functions.* National Bureau of Standards, Washington D.C., 1964.

[BI(BII)] Bary, N. *A Treatise on Trigonometric Series* I (II). Macmillan, New York, 1964.

[BH] Berglund J. and Hofmann, K. *Compact Semitopological Semigroups and Weakly Almost Periodic Functions.* Springer-Verlag, Berlin, 1967.

[B] Boerner, H. *Representations of Groups,* North-Holland, Amsterdam, 1963.

[C] Calderón, A. *Integrales singulares y sus aplicaciones a ecuaciones diferenciales hiperbolicas.* Universidad de Buenos Aires, Buenos Aires, 1960.

[DS] Dunford, N. and Schwartz, J. *Linear Operators,* Part I. Interscience Publishers, New York, 1958.

[E] Edwards, R. *Fourier Series* II. Holt, New York, 1967.

[H] Halmos, P. *Measure Theory.* D. Van Nostrand, Princeton, 1950.

[HR] Hewitt, E. and Ross, K. *Abstract Harmonic Analysis* I. Academic Press, New York, 1963.

[Hö] Hörmander, L. *An Introduction to Complex Analysis in Several Variables.* D. Van Nostrand, Princeton, 1966.

[K] Kahane, J.-P. *Some Random Series of Functions.* Heath, Lexington, Massachusetts, 1968.

[Ka] Kaplansky, I. *Infinite Abelian Groups.* University of Michigan Press, Ann Arbor, 1954.

[KN] Kelly, J. and Namioka, I. *Linear Topological Spaces.* D. Van Nostrand, Princeton, 1963.

[Ko] Koethe, G. *Topologische Lineare Räume* I. Springer-Verlag, Berlin, 1960.

[Ku] Kurosh, A. *The Theory of Groups* I. Chelsea, New York, 1955.

[L] Loomis, L. *An Introduction to Abstract Harmonic Analysis.* D. Van Nostrand, Princeton, 1953.

[N] Naimark, M. *Normed Rings*. P. Noordhoff N.V., Groningen, The Netherlands, 1964.

[Ra] Radó, T. *Subharmonic Functions*. Springer-Verlag, Berlin, 1937.

[Ri] Rickart, C. *General Theory of Banach Algebras*. D. Van Nostrand, Princeton, 1960.

[R] Rudin, W. *Fourier Analysis on Groups*. Interscience Publishers, New York, 1962.

[RP] Rudin, W. *Principles of Mathematical Analysis*. McGraw-Hill, New York, 1964.

[RRC] Rudin, W. *Real and Complex Analysis*. McGraw-Hill, New York, 1966.

[RFT] Rudin, W. *Function Theory in Polydiscs*. Benjamin, New York, 1969.

[W] Weil, A. *L'intégration dans les Groupes Topotogtques et ses Applications, Actualites Sci. et Ind.* 869–1145. Hermann, Paris, 1938.

[Y] Yosida, K. *Functional Analysis*, 2nd ed. Springer-Verlag, Berlin, 1968.

[Z] Zygmund, A. *Trigonometric Series* I. Cambridge University Press, New York, 1959.

Research Publications

Amemiya, I. and Itô, T.

[1] A simple proof of the theorem of P. J. Cohen. *Bull. Amer. Math. Soc.* **70** (1964), 774–776.

Banach, S.

[1] Über einige Eigenschaften der lakunären trigonometrischen Reihen. *Studia Math.* **2** (1930), 207–220.

Bredon, G.

[1] A new treatment of the Haar integral. *Michigan Math. J.* **10** (1963), 365–373.

Buck, R. C.

[1] Operator algebras and dual spaces. *Proc. Amer. Math. Soc.* **3** (1952), 681–687.

Bungart, L.

[1] Boundary kernel functions for domains on complex manifolds. *Pacific J. Math.* **14** (1964), 1151–1164.

Cartan, E.

[1] Sur la determination d'un système orthogonal complet dans un espace de Riemann symmetrique clos, *Rend. Circ. Mat. Palermo* **53** (1929), 217–252.

Chaney, R.

[1] On uniformly approximable Sidon sets. *Proc. Amer. Math. Soc.* **21** (1969), 245–249.

Cohen, P.

[1] On a conjecture of Littlewood and idempotent measures. *Amer. J. Math.* **82** (1960), 191–212.

[2] On hormomorphisms of group algebras. *Amer. J. Math.* **82** (1960), 213–226.

De Leeuw, K.

[1] Functions on circular subsets of the space of n complex variables. *Duke Math. J.* **24** (1957), 415–432.

De Leeuw, K. and Glicksberg, I.

[1] Applications of almost periodic compactifications. *Acta Math.* **105** (1961), 63–97.

[2] Almost periodic functions on semigroups. *Acta Math.* **105** (1961), 99–140.

[3] The decomposition of certain group representations. *J. Analyse Math.* **15** (1965), 135–192.

Drury, S.

[1] Sur les ensembles de Sidon. *C.R. Acad. Sc. Paris* 271. Séries A (1970), 162–163.

Dunkl, C.

[1] Operators and harmonic analysis on the sphere. *Trans. Amer. Math. Soc.* **125** (1966), 250–263.

[2] Linear functionals on homogeneous polynomials. *Canad. Math. Bull.* **11** (1968), 465–468.

[3] Functions that operate in the Fourier algebra of a compact group. *Proc. Amer. Math. Soc.* **21** (1969), 540–544.

[4] Modules over commutative Banach algebras. *Monatsh. Math.* **74** (1970), 6–14.

Dunkl, C. and Ramirez, D.

[1] Translation in measure algebras and the correspondence to Fourier transforms vanishing at infinity. *Michigan Math. J.* **17** (1970), 311–319.

[2] Homomorphisms on groups and induced maps on certain algebras of measures. *Trans. Amer. Math. Soc.* (to appear).

[3] Helson sets in compact and locally compact groups *Michigan Math. J.* (to appear).

[4] Sidon sets on compact groups. *Monatsh. Math.* (to appear).

Eberlein, W.

[1] Abstract ergodic theorems and weak almost periodic functions. *Trans. Amer. Math. Soc.* **67** (1949), 217–240.

[2] The point spectrum of weakly almost periodic functions. *Michigan Math. J.* **3** (1955–56), 137–139.

Edwards, R.

[1] On functions which are Fourier transforms. *Proc. Amer. Math. Soc.* **5** (1954), 71–78.

[2] Uniform approximation on noncompact spaces. *Trans. Amer. Math. Soc.* **122** (1966), 249–276.

Edwards, R. and Hewitt, E.

[1] Pointwise limits for sequences of convolution operators. *Acta Math.* **133** (1965), 181–217.

Eymard, P.

[1] L'algèbre de Fourier d'un groupe localement compact. *Bull. Soc. Math. France* **92** (1964), 181–236.

Figà-Talamanca, A. and Rider, D.

[1] A theorem of Littlewood and lacunary series for compact groups. *Pacific J. Math.* **16** (1966), 505–514.

[2] A theorem on random Fourier series on compact groups. *Pacific J. Math.* **21** (1967), 487–492.

Gårding, L. and Hörmander, L.

[1] Strongly subharmonic functions. *Math. Scand.* **15** (1964), 93–96.

Glicksberg, I. (see De Leeuw)

Godement, R.

[1] A theory of spherical functions I. *Trans. Amer. Math. Soc.* **73** (1952), 496 556.

Grothendieck, A.

[1] Critères de compacité dans les espaces fonctionnels genéraux. *Amer. J. Math.* **74** (1952), 168–186.

Haar, A.

[1] Der Massbegriff in der Theorie der kontinuierlichen Gruppen. *Ann. Math.* **34** (1933), 147–169.

Halmos, P. and Vaughan, H.

[1] The marriage problem. *Amer. J. Math.* **72** (1950), 214–215.

Helgason, S.

[1] Topologies of group algebras and a theorem of Littlewood. *Trans. Amer. Math. Soc.* **86** (1957), 269–283.

[2] Lacunary Fourier series on noncommutative groups. *Proc. Amer. Math. Soc.* **9** (1958), 782–790.

Helson, H.

[1] Note on harmonic functions. *Proc. Amer. Math. Soc.* **4** (1953), 686–691.

[2] On a theorem of Szegö. *Proc. Amer. Math. Soc.* **6** (1955), 235–242.

Helson, H., Kahane, J.-P., Katznelson, Y., and Rudin W.

[1] The functions which operate on Fourier transforms. *Acta Math.* **102** (1959), 135–157.

Hewitt, E. (see Edwarde).

[1] Representation of functions as absolutely convergent Fourier-Stieltjes transforms. *Proc. Amer. Math. Soc.* **4** (1953), 663–670.

[2] A survey of abstract harmonic analysis. Some Aspects of Analysis and Probability, pp. 105–168. *Surveys in Applied Mathematics,* vol. 4. John Wiley, New York, 1958.

Hewitt, E. and Zuckerman, H.

[1] Some theorems on lacunary Fourier series, with extensions to compact groups. *Trans. Amer. Math. Soc.* **93** (1959), 1–19.

Hörmander, L. (see Gårding)

Itô, T. (see Amemiya)

Johnson, B.

[1] Isometric isomorphisms of measure algebras. *Proc. Amer. Math. Soc.* **15** (1964), 186–188.

[2] Symmetric maximal ideals in $M(G)$. *Proc. Amer. Math. Soc.* **18** (1967), 1040–1044.

[3] The Shilov boundary of $M(G)$. *Trans. Amer. Math. Soc.* **134** (1968), 289–296.

Kahane, J.-P. (see Helson)

[1] Sur les fonctions moyenne -périodiques bornées. *Annales Inst. Fourier.* **7** (1957), 293–314.

Katznelson, Y. (see Helson)

Keogh, F.

[1] Riesz products. *Proc. London Math. Soc.* **3** (1965), 174–182.

Kessler, I.

[1] Semi-idempotent measures on abelian groups. *Bull. Amer. Math. Soc.* **73** (1967), 258–260.

Krein, M.

[1] Hermitian-positive kernels on homogeneous spaces I. *Ukrain. Mat. Žurnal* **1** (1949), 64–98. *Amer. Math. Soc. Translations,* Series 2, Vol. **34,** 69–108.

[2] Hermitian-positive kernels on homogeneous spaces II. *Ukrain. Mat. Žurnal* **2** (1950), 10–59. *Amer. Math. Soc. Translations,* Series 2, Vol. **34,** 109–164.

Kunze, R.

[1] L_p Fourier transforms on locally compact unimodular groups. *Trans. Amer. Math. Soc.* **89** (1958), 519–540.

Mayer, R.

[1] Summation of Fourier series on compact groups. *Amer. J. Math.* **89** (1967), 661–692.

[2] Fourier series of differentiable functions on $SU(2)$. *Duke Math. J.* **34** (1967), 549–554.

[3] Localization for Fourier series on $SU(2)$. *Trans. Amer. Math. Soc.* **130** (1968), 414–424.

[4] An example of nonlocalization for Fourier series on $SU(2)$. *Illinois J. Math.* **12** (1968), 325–334.

Miller, R.
[1] Gleason parts and Choquet boundary points in convolution measure algebras. *Pacific J. Math.* **32** (1969), 755–771.

Peter, F. and Weyl, H.
[1] Die Vollständigkeit der primitiven Darstellungen einer geschlossenen kontinuierlichen Gruppe. *Math. Annalen* **97** (1927), 737–755.

Pym, J.
[1] The convolution of functionals on spaces of bounded functions. *Proc. London Math. Soc.* **15** (1965), 84–104.

Ramirez, D. (see Dunkl)
[1] Uniform approximation by Fourier-Stieltjes transforms. *Proc. Cambridge Philos. Soc.* **64** (1968), 323–333.
[2] Uniform approximation by Fourier-Stieltjes coefficients. *Proc. Cambridge Philos. Soc.* **64** (1968), 615–623.
[3] Weakly almost periodic function and Fourier-Stieltjes transforms. *Proc. Amer. Math. Soc.* **19** (1968), 1087–1088.
[4] The measure algebra as an operator algebra. *Canad. J. Math.* **20** (1968), 1391–1396.
[5] A characterization of quotient algebras of $L^1(G)$. *Proc. Cambridge Philos. Soc.* **66** (1969), 547–551.

Rennison, J.
[1] Arens products and measure algebras. *J. London Math. Soc.* **44** (1969), 369–377.

Rider, D. (see Figà-Talamanca)
[1] Function algebras on groups and spheres. *Notices Amer. Math. Soc.* **11** (1964), 544.
[2] Gap series on groups and spheres. *Canad J. Math.* **18** (1966), 389–398.
[3] Central idempotent measures on unitary groups. *Canad. J. Math.* **22** (1970)
[4] Central idempotent measures on *SIN* groups. *Duke Math. J.* **38** (1971)
[5] Functions which operate in the Fourier algebra of a compact group (to appear).

Rudin, W. (see Helson)
[1] Idempotent measures on abelian groups. *Pacific J. Math.* **9** (1959), 195–209.
[2] Weak almost periodic functions and Fourier-Stieltjes transforms. *Proc. Amer. Math. Soc.* **26** (1959), 215–220.
[3] Measure algebras on abelian groups. *Bull. Amer. Math. Soc.* **65** (1959), 227–247.
[4] Trigonometric series with gaps. *J. Math. Mech.* **9** (1960), 203–228.
[5] Projections on invariant subspaces. *Proc. Amer. Math. Soc.* **13** (1962), 429–432.

Saeki, S.
[1] On norms of idempotent measures. *Proc. Amer. Math. Soc.* **19** (1968), 600–602.
[2] On norms of idempotent measures II. *Proc. Amer. Math. Soc.* **19** (1968), 367–371.

Saitô, K.
[1] On a duality for locally compact groups. *Tohoku Math. J.* **20** (1968), 355–367.

Segal, I.
[1] The class of functions which are absolutely convergent Fourier transforms. *Acta Sci. Math. Szeged* **12** (1950), 157–161.
[2] An extension of Plancherel's formula to separable unimodular groups. *Ann. Math.* **52** (1950), 272–292.
[3] A noncommutative extension of abstract integration. *Ann. Math.* **57** (1953), 401–457.

Shiga, K.
[1] Representations of a compact group on a Banach space. *J. Math. Soc. Japan* **7** (1955), 224–248.
Sidon, S.
[1] Über orthogonale Entwicklungen. *Acta Univ. Szeged.* **10** (1941–1943), 206–253.
Simon, A.
[1] Symmetry in measure algebras. *Bull. Amer. Math. Soc.* **66** (1960), 399–400.
Šreider, Yu.
[1] The structure of maximal ideals in rings of measures with convolution. *Amer. Math. Soc. Transl.* (1) **8** (1962), 365–391.
Stečken, S.
[1] On absolute convergence of Fourier series. *Izv. Akad. Nauk SSSR, Ser. Mat.* **20** (1956), 385–412.
Stinespring, W.
[1] Integration theorems for gages and duality for unimodular groups. *Trans. Amer. Math. Soc.* **90** (1959), 15–56.
Strichartz, R.
[1] Isometric isomorphisms of measure algebras. *Pacific J. Math.* **15** (1965), 315–317.
Taylor, J.
[1] The structure of convolution measure algebras. *Trans. Amer. Math. Soc.* **119** (1965), 150–166.
[2] The Shilov boundary of the algebra of measures on a group. *Proc. Amer. Math. Soc.* **16** (1965), 941–945.
Taylor, M.
[1] Fourier series on compact Lie groups. *Proc. Amer. Math. Soc.* **19** (1968), 1103–1105.
Varopoulous, N.
[1] Measure algebras of a locally compact abelian group. *Séminaire Bourbaki:* Vol. 1964/65 Expose 282. Benjamin, New York, 1966.
Vaughan, H. (see Halmos)
Wendel, J. G.
[1] On isometric isomorphisms of group algebras. *Pacific J. Math.* **1** (1951), 305–311.
Weyl, H. (see Peter)
Williamson, J.
[1] A theorem on algebras of measures on topological groups. *Proc. Edinburgh Math. Soc.* **11** (1958–1959), 195–206.
Zuckerman, H. (see Hewitt)

INDEX OF SPECIAL SYMBOLS

AUTHOR INDEX

SUBJECT INDEX